Toxicology and Environmental Health Information Resources
The Role of the National Library of Medicine

Catharyn T. Liverman, Carrie E. Ingalls, Carolyn E. Fulco, and Howard M. Kipen, *Editors*

Committee on Toxicology and Environmental Health Information Resources for Health Professionals

Division of Health Promotion and Disease Prevention

INSTITUTE OF MEDICINE

NATIONAL ACADEMY PRESS
Washington, D.C. 1997

NATIONAL ACADEMY PRESS • 2101 Constitution Avenue, N.W. • Washington, DC 20418

NOTICE: The project that is the subject of this report was approved by the Governing Board of the National Research Council, whose members are drawn from the councils of the National Academy of Sciences, the National Academy of Engineering, and the Institute of Medicine. The members of the committee responsible for the report were chosen for their special competencies and with regard for appropriate balance.

This report has been reviewed by a group other than the authors according to procedures approved by a Report Review Committee consisting of members of the National Academy of Sciences, the National Academy of Engineering, and the Institute of Medicine.

The Institute of Medicine was chartered in 1970 by the National Academy of Sciences to enlist distinguished members of the appropriate professions in the examination of policy matters pertaining to the health of the public. In this, the Institute acts under both the Academy's 1863 congressional charter responsibility to be an advisor to the federal government and its own initiative in identifying issues of medical care, research, and education. Dr. Kenneth I. Shine is president of the Institute of Medicine.

Support for this project was provided by the National Library of Medicine, National Institutes of Health, under Contract No. N01-OD-4-2139. The views presented are those of the Institute of Medicine Committee on Toxicology and Environmental Health Information Resources for Health Professionals and are not necessarily those of the funding organization.

International Standard Book No. 0-309-05686-1

Additional copies of this report are available for sale from the National Academy Press, Box 285, 2101 Constitution Avenue, Washington, DC 20055. Call (800) 624-6242 or (202) 334-3313 in the Washington metropolitan area, or visit the NAP's on-line bookstore at **http://www.nap.edu**.

Copyright 1997 by the National Academy of Sciences. All rights reserved.

Printed in the United States of America.

COMMITTEE ON TOXICOLOGY AND ENVIRONMENTAL HEALTH INFORMATION RESOURCES FOR HEALTH PROFESSIONALS

HOWARD KIPEN, (*Chair*), Associate Professor and Director, Division of Occupational Health, Environmental and Occupational Health Sciences Institute, University of Medicine and Dentistry of New Jersey—Robert Wood Johnson Medical School, Piscataway, New Jersey

PAUL FRAME,[*] Family Physician, Tri-County Family Medicine, Cohocton, New York

MARK FRISSE, Associate Dean, Washington University School of Medicine, St. Louis, Missouri

SHERRILYNNE FULLER, Acting Director, Informatics, School of Medicine, and Director, Health Sciences Libraries and Information Center, University of Washington, Seattle, Washington

FRED HENRETIG, Pediatric Emergency Physician, Children's Hospital of Philadelphia, Philadelphia, Pennsylvania

DAVID McNELIS, Chief Scientist, Environmental Science and Engineering, Research Triangle Institute, Research Triangle Park, North Carolina

KATHLEEN REST, Associate Professor, Department of Family and Community Medicine, Occupational and Environmental Health Program, University of Massachusetts School of Medicine, Worcester, Massachusetts

BARBARA SATTLER, Director, Environmental Health Education Center, University of Maryland at Baltimore School of Medicine, Baltimore, Maryland

ROSE ANN SOLOWAY, Administrator, American Association of Poison Control Centers, and Clinical Toxicologist, National Capital Poison Center, Washington, D.C.

ROBERT E. TAYLOR, Chairman, Department of Pharmacology, Howard University College of Medicine, Washington, D.C.

P. IMANI THOMPSON, Behavioral Scientist, Centers for Disease Control and Prevention, Atlanta, Georgia

Institute of Medicine Staff
CATHARYN T. LIVERMAN, Study Director
CAROLYN E. FULCO, Senior Program Officer
CARRIE E. INGALLS, Research Associate
THOMAS WETTERHAN, Administrative Assistant/Research Assistant
AMELIA MATHIS, Project Assistant
MICHAEL STOTO, Director, Division of Health Promotion and Disease Prevention (through December 1996)
KATHLEEN STRATTON, Interim Director, Division of Health Promotion and Disease Prevention (from January 1997)

---

[*] Member, Institute of Medicine.

Preface

The environment is increasingly recognized as having an impact on human and ecological health, as well as on specific types of human morbidity, mortality, and disability. Since the publication of its landmark report in 1988, *Role of the Primary Care Physician in Occupational and Environmental Medicine*, the Institute of Medicine (IOM) has conducted two additional studies that have examined the need to integrate environmental and occupational health into the education and practices of nurses and physicians. The recommendations from these reports are currently being implemented.

In 1995, the National Library of Medicine (NLM) asked the IOM to explore a related topic by requesting a study of NLM's Toxicology and Environmental Health Information Program (TEHIP). Specifically, NLM was concerned that health professionals were not fully using the information available in the 16 online databases comprising the TEHIP program. The IOM formed the Committee on Toxicology and Environmental Health Information Resources for Health Professionals. One of the committee's first goals was to seek input from a wide range of health professionals to more thoroughly understand health professionals' toxicology and environmental health information needs. Several mechanisms were used to receive input, including a workshop, during which attendees participated in focus group sessions; a questionnaire, designed to solicit information about health professionals' information needs; and discussions with representatives from federal agencies, health care, and academia.

During the course of the study, the committee reached several conclusions that it viewed as pivotal in advising NLM on how best to provide health professionals with toxicology and environmental health information. First, the committee believes that as environmental health concerns continue to increase, it is important for health professionals and other communities to have ready access to information resources in this field. The committee reaffirms the findings of the 1993 NLM Long Range Planning Panel on Toxicology and Environmental Health, which found that NLM's TEHIP program is an important information resource that needs to be strengthened.

Second, the committee believes that there is a large and diverse potential audience for toxicology and environmental health information. In attempting to understand the user communities, the committee discussed a broad spectrum of potential users ranging from emergency care personnel treating individuals affected by acute toxic exposures to local coalitions struggling to determine the environmental health hazards faced by their communities. Although each of the user communities in this broad spectrum has diverse information needs, there are methods of targeting training and outreach efforts and developing database interfaces that will more adequately meet those disparate needs.

Finally, the committee concluded that NLM, as the nation's premier biomedical library, can and should play a key role in organizing and providing pointers to all toxicology and environmental health information resources (including and beyond the TEHIP databases). NLM, given its library and medical informatics expertise, is well-positioned to further develop the tools that can link health professionals with the wide array of information resources that are available in this important field. Furthermore, this is an area where public-private sector partnerships can play an important role as there are numerous sources of toxicology and environmental health information.

The committee is grateful to those who provided input to its deliberations including the individuals who contributed their ideas through the workshop and in discussions with the committee (see Appendixes A and C). Additionally, the committee thanks the individuals who took the time to respond to the committee's questionnaire (see Appendix B). The TEHIP program staff, including Jeanne Goshorn and Melvin Spann, provided thorough background materials, assisted in the committee's workshop, and responded promptly to the committee's many requests for additional information or clarification. The committee appreciates all of their efforts. The IOM staff of Cathy Liverman, Carrie Ingalls, and Carolyn Fulco are to be congratulated for their thorough research of the issues and for molding the committee's sometimes wandering deliberations into this report.

Although the committee has recommended some clear directions and mechanisms for implementation, much work remains to be done. The committee hopes that the conclusions and recommendations made in this report will prove to be useful as NLM moves forward in providing health professionals with toxicology and environmental health information.

                  Howard M. Kipen, M.D., *Chair*
                  Committee on Toxicology and Environmental Health
                  Information Resources for Health Professionals

Contents

EXECUTIVE SUMMARY 1

1 INTRODUCTION 11
  Health Professionals and Other User Communities, 13
  Public Health Impacts of Hazardous Substances, 14
  Changing Trends in Health Practice, 15
  Organization of the Report, 16

**2 THE NATIONAL LIBRARY OF MEDICINE'S TOXICOLOGY AND
    ENVIRONMENTAL HEALTH INFORMATION PROGRAM** 19
  National Library of Medicine, 20
  Division of Specialized Information Services, 25
  Toxicology and Environmental Health Information Program, 26
  TEHIP Databases, 29
  TEHIP Factual Databases, 30
  TEHIP Bibliographic Databases, 47
  Conclusions, 52

**3 OTHER TOXICOLOGY AND ENVIRONMENTAL HEALTH
    INFORMATION RESOURCES** 55
  Conclusion and Recommendation, 58

4 UNDERSTANDING THE INFORMATION NEEDS OF HEALTH PROFESSIONALS 69
Information Needs, 69
Factors Affecting Information Seeking, 71
Current and Potential Users of the TEHIP Databases, 75
Conclusion and Recommendation, 82

5 INCREASING AWARENESS: TRAINING AND OUTREACH 87
Training, 88
Outreach, 91
Future Directions and Recommendation, 96

6 ACCESSING AND NAVIGATING THE TEHIP DATABASES 101
Access: Getting Connected to the Databases, 101
Navigating the TEHIP Databases, 105
Future Directions, 114
Conclusions and Recommendations, 116

7 PROGRAM ISSUES AND FUTURE DIRECTIONS 119
Program Issues, 120
Future Directions, 126

GLOSSARY AND ACRONYMS 131

APPENDIXES
A   Acknowledgments, 143
B   Questionnaire, 145
C   Workshop on Toxicology and Environmental Health Information Resources: Agenda, Participants, and Summary of Focus Group Discussions, 153

TABLES, FIGURES, AND BOXES

Tables

2.1 Timeline of Events and Changes in Computer Technology and Environmental Health, 23
2.2 TEHIP Databases, 31
2.3 Types of Information Available in the TEHIP Databases, 35
2.4 TOXLINE Subfiles, 48
3.1 Sample of Current Toxicology and Environmental Health Databases, 59
6.1 Review Process for TEHIP Factual Databases, 113

Figures

2.1 National Library of Medicine Organizational Chart, 21
2.2 TEHIP Program Budget, 28
2.3 Organization of the TEHIP Databases, 29
3.1 Executive Branch Departments and Agencies Involved in Environmental Health Issues, 57
5.1 National Network of Libraries of Medicine, 93
6.1 Primary Access Points to the TEHIP Databases, 104
7.1 NLM Advisory Committees, 122
B.1 Toxicology and Environmental Health Information Resources Most Often Consulted, 147
B.2 Primary Factors Limiting Use of the NLM Toxicology and Environmental Health Databases, 147

Boxes

2.1 Locator Field in the ChemID Database, 36
2.2 Major Categories of HSDB Data, 37
2.3 Excerpt from the TRIFACTS Record on Toluene, 45
2.4 DIRLINE Record for the Association of Occupational and Environmental Clinics, 46
4.1 Examples of the Applicability of the TEHIP Databases for the Work of Health Professionals, 78
5.1 Previous IOM Recommendations on the Training of Health Professionals in Occupational and Environmental Health, 90
5.2 Howard University, 95
6.1 Methods of Searching the TEHIP Databases, 102
6.2 TOXNET Selection Menu, 106

6.3 Sample Search on HSDB, 109
6.4 Initial TRI Menu, 110
6.5 Experimental World Wide Web Search Interface: Criteria for Narrowing the Search Strategy, 111

Toxicology and Environmental Health Information Resources: The Role of the National Library of Medicine

Executive Summary

Health professionals need access to environmental health[1] and toxicology information for many reasons. Certainly, public awareness about human health risks from chemical and biologic agents in the environment has increased dramatically in recent years. Similarly, changing trends in health care and an emphasis on prevention, coupled with increasing computer literacy, all support the need for readily available information about the impacts of hazardous substances in the environment on individual and public health. Reports in the popular press and news media have highlighted the public's concern. For example, pesticides on foods; second-hand tobacco smoke; asbestos and lead paint in homes and public buildings; dioxin contamination; occupational exposures to gasoline and other chemicals; exposure to radon and benzene; and drinking water contaminated with biologic or chemical agents are just a few of the issues that may confront the American public.

Although the public relies heavily on federal and state regulatory agencies for protection from exposures to hazardous substances, they frequently look to health professionals for information on routes of exposure and the nature and extent of associated adverse health consequences. However, most health professionals acquire only a minimal knowledge of toxicology during their education and training. As a result, their working knowledge of the adverse effects of chemicals on health and the conditions under which those effects might occur is often limited. Furthermore, with the many competing demands on health professionals' time, it is difficult, even for specialists, to keep apprised of rapidly

---

[1] The committee's use of the term *environmental health* includes health issues involving exposures to hazardous substances in the workplace, home, and community settings.

evolving toxicology information. Thus, health professionals need ready access to toxicology and environmental health information resources to assist them with patient care. Policymakers, health advisors, researchers, health educators, and the general public also need access to this information as they pursue their own inquiries.

Established in 1956, the National Library of Medicine (NLM) was charged with improving the nation's health by collecting and providing access to the world's biomedical literature. In 1967, NLM established a specialized information program on toxicology and environmental health known today as the Toxicology and Environmental Health Information Program (TEHIP). Its mission is to provide selected core information resources and services, facilitate access to national and international information resources, and strengthen the infrastructure for toxicology and environmental health information.

In 1995, at the request of NLM, the Institute of Medicine (IOM) formed the Committee on Toxicology and Environmental Health Information Resources for Health Professionals. The committee was charged with examining the utility and accessibility[2] of NLM's TEHIP program for the work of health professionals. It was asked to consider the current toxicology and environmental health information needs of health professionals and how those needs are currently being met. The committee conducted an 18-month study and received extensive input from health professionals representing a range of disciplines and expertise. These individuals met with the committee, participated in one of four focus group sessions at a committee workshop, or responded to a questionnaire developed by the committee. Additionally, NLM staff members provided the committee with information about the databases and ongoing research and development efforts at NLM.

CONTEXT OF THE REPORT

Both the public health effects of hazardous substances and the changing trends in health care are reinforcing the need for authoritative and easily accessible information in the fields of toxicology and environmental health. Pathways of human exposure to chemicals in the environment are diverse. Chemicals in air, food, water, and soil can be inhaled, ingested, or absorbed in any number of settings. Vulnerable populations, such as children, the elderly, the chronically ill, minorities, and the poor may be at increased risk of harm related to environmental contamination because of biologic and demographic factors, including where they live.

---

[2]Factors examined in assessing the utility (usefulness) and accessibility (ease of use) of the databases include the subject content, search interface, and available access points.

Despite the complexities involved in studying human exposures, the roles of certain hazardous substances in the development of human disease are well-known. Environmental and occupational chemical exposures can affect all organ systems. They can cause or contribute to the development of a variety of human illness, including cancer, asthma and other respiratory diseases, reproductive disorders, neurologic and immune system impairments, and skin disease, as well as cardiovascular, renal, hepatic, and psychological disorders. The proliferation of the manufacture, transport, use, and disposal of chemicals coupled with the potential for human exposures makes it essential that health professionals have easy access to resources and information to assist them in the prevention, diagnosis, and treatment of disorders possibly stemming from exposures to hazardous substances in the environment.

NLM'S TOXICOLOGY AND ENVIRONMENTAL HEALTH INFORMATION PROGRAM

Currently, the TEHIP program encompasses 16 online databases that contain bibliographic and factual information on environmental contaminants including chemical properties, carcinogenicity, exposure levels, adverse health effects, emergency treatment protocols, and federal regulations. The evolution of NLM's TEHIP program has been the result of both internal NLM commitments to developing toxicology and environmental health information resources (e.g., TOXLINE and the Hazardous Substances Data Bank) and the interests of other federal agencies in fulfilling their missions and legislative mandates (e.g., the Environmental Protection Agency's Toxic Chemical Release Inventory).

Although the budgeted funding for the TEHIP program (from federal appropriations) has remained relatively constant over the past 29 years (in fiscal year [FY] 1994, the TEHIP program's appropriated budget in current dollars was approximately $7.4 million), the TEHIP program's reimbursements from other agencies for collaborative projects have fluctuated. In FY 1992, the TEHIP program's total reimbursable budget from other agencies was $2.45 million, whereas in FY 1993 the reimbursable budget dropped by approximately 50 percent to $1.27 million. Since 1993, the reimbursable budget has remained relatively level (the FY 1995 reimbursable budget was $1.23 million).

OTHER INFORMATION RESOURCES

The TEHIP databases represent only a small subset of the numerous databases containing information related to toxicology and environmental health. Responsibilities for research, regulation, and risk communication on environmental health issues are fragmented between numerous local, state, and federal government agencies, international organizations, industry, and other private-sector businesses. As a result, numerous databases that are relevant to fulfilling each organization's specific environmental health mission or goals have been developed. The committee became aware of the vast extent of information available and the utility of these information resources to health professionals. The committee believes that it is important for NLM to carry out the traditional and expert role of a library by organizing (cataloging) online information resources in toxicology and environmental health beyond the TEHIP databases and increasing health professionals' and other interested users' awareness of the relevant resources. By providing users with information on non-NLM resources (e.g., a description of the information resource and its access points), NLM will be delivering the valuable library service of providing users with the information needed to access the most relevant resource available.

The committee recommends that NLM consider expanding its traditional library services in toxicology and environmental health by organizing and cataloging the full spectrum of online toxicology and environmental health information resources.

UNDERSTANDING THE INFORMATION NEEDS OF HEALTH PROFESSIONALS

As knowledge about the health effects of exposure to occupational and environmental chemicals increases, health professionals and other interested individuals need to be able to access and use resources that provide timely and accurate toxicology and environmental health information in an efficient and accessible manner. The committee discussed the health professional communities that potentially have information needs in toxicology and environmental health and use for the information available in the TEHIP databases. The committee realizes that the health professional community does not have homogeneous information needs and that there is a wide range of variation in access to online databases and other information resources. Potential user communities include primary care professionals and pharmacists; specialists in occupational and environmental health; emergency medicine and poison control center personnel; health science librarians and faculty at health professional schools; environmental health researchers and scientists; patients, the general public, and community organizations; and health professionals in local public health departments or in state and federal agencies.

EXECUTIVE SUMMARY

Through committee deliberations and input provided to the committee from the focus groups and questionnaire respondents,[3] the disparate nature of the 16 databases in the TEHIP program became more evident. The committee recognized that the databases are not equally useful for the work of the different segments of the health professional community. Although the committee realized that the current TEHIP complement of databases is the result of both NLM initiatives and interagency agreements and that each database fills an important information niche, the committee believes that the TEHIP program should set priorities that would allow efforts to be focused on those databases that meet the information needs of the greatest number of health professionals. This is particularly critical in light of the fact that the TEHIP program has experienced reduced funding levels from interagency agreements in recent years.

Comparable to a business marketing strategy that necessitates an understanding of the specific needs of current and potential customers prior to designing and distributing the product, this prioritization of the TEHIP program would first require a more in-depth analysis of the toxicology and environmental health information needs of health professionals. The goal of this user profile analysis would be to match, as closely as possible, the needs of health professionals with specific TEHIP databases. Upon completion of the user analysis, TEHIP program staff could not only prioritize their training and outreach efforts with an emphasis on those databases that are the most useful to health professionals but could also prioritize the resources that are devoted to the databases with the greatest utility for health professionals.

The committee recommends that NLM further expand its efforts to understand the toxicology and environmental health information needs of health professionals and the barriers they face in accessing that information by conducting a detailed user profile analysis. Additionally, the committee recommends that the results from that analysis be used to set priorities for subsequent efforts of the TEHIP program.

---

[3] It is important to note the limitations of the focus groups and questionnaire. The committee did not attempt to obtain a random scientific sample for the distribution of the questionnaire. Rather, the questionnaire was distributed both to professional association members and via the Internet. Thus, it was not feasible to determine the rates of response or to characterize the nonresponders. Additionally, responding via the Internet requires some degree of computer expertise. Because travel time and expenses were considerations in inviting focus group participants, most of the participants were drawn from the mid-Atlantic, particularly the Washington, D.C. metropolitan area.

INCREASING AWARENESS: TRAINING AND OUTREACH

An important component of expanding the use of toxicology and environmental health information resources is increasing the awareness of these resources in the health professional and other interested user communities. Potential users of online toxicology and environmental health databases must be cognizant of the existence of the databases and of their content, must be computer literate (assuming that the user will perform his or her own search), and must have some familiarity with toxicology and environmental health data in order to interpret the retrieved information correctly.

Computer use is largely a matter of demographics. Younger health professionals are more likely to feel comfortable with computer use and to have become accustomed to retrieving information through computers. There is, however, a continuing need to train health professionals about specific databases and the use of health-related information resources.

Although no health professional can be expected to know the toxic effects of all chemicals, it is critical that health professionals be informed about the issues and familiar enough with the field to consider environmental and occupational exposures in assessing a patient's symptoms, making a diagnosis, answering a patient's questions, and counseling patients about environmental health risks. However, health professionals, especially clinicians, receive limited education and training in toxicology and environmental health in part because of an overcrowded and increasingly specialized curriculum. The committee supports the recommendations of several recent IOM studies (*Environmental Medicine: Integrating a Missing Element into Medical Education* [1995]; *Nursing, Health, and the Environment: Strengthening the Relationship to Improve the Public's Health* [1995]) that have focused on strategies for enhancing the environmental health content in health sciences curriculum and continuing education courses.

NLM's training and outreach efforts are primarily conducted through the National Network of Libraries of Medicine (NN/LM), a nationwide network of more than 4,500 local medical libraries (primarily hospital libraries), more than 140 Resource Libraries (primarily at medical schools), and eight Regional Medical Libraries. The medical librarians and information specialists at each NN/LM member library play an integral role in educating, training, and providing access to NLM's resources.

Additionally, the Division of Specialized Information Services (SIS) (which manages the TEHIP program) has been active in targeting outreach efforts to reach those populations that are particularly interested in environmental health information. In 1991, SIS implemented a pilot Toxicology Information Outreach Project with the objective of strengthening the capacity of Historically Black Colleges and Universities to train medical and other health professionals in the use of NLM's toxicology and environmental health information resources. Additionally, SIS is actively involved in a collaborative project between NLM and the Wheaton Regional Public Library (in Montgomery County, Maryland) that promotes access to information resources on environmental health and HIV/AIDS.

The committee believes that the TEHIP program can make significant contributions to health professionals by providing information on the wide scope of relevant toxicology and environmental health information resources. Outreach efforts should be targeted to meet the specific interests of the audience and should be expanded to utilize dissemination networks and environmental health efforts and initiatives currently underway. Utilization of the user analysis could assist NLM in focusing its outreach and training efforts on those databases that best meet the needs of the user community.

The committee recommends that NLM's training and outreach efforts in toxicology and environmental health information be increased to improve awareness and recognition of these resources. Mechanisms that may improve awareness include emphasizing the broad spectrum of toxicology and environmental health information resources, matching training to meet the specific needs of the target audiences, and expanding the use of already-existing distribution mechanisms for promoting the availability of toxicology and environmental health information.

ACCESSING AND NAVIGATING THE TEHIP DATABASES

Access to the TEHIP databases and difficulties in navigating the user interface were selected by focus group participants and those individuals responding to a committee questionnaire as two of the primary factors limiting use of the TEHIP databases. The only fully operational access method available for accessing all of the TEHIP databases is direct searching, a complex command-line method requiring considerable expertise in online searching. Grateful Med—a software program developed by NLM—provides a user-friendly interface for searching the NLM databases, however Grateful Med does not offer input screens for all of the TEHIP databases. NLM staff are in the process of developing a new approach to searching the TEHIP databases using the hypertext and graphics capabilities available through the World Wide Web.

The committee believes that to increase the use of the TEHIP databases by health professionals, it is critically important that they be easily accessible and that the user interfaces be intuitive. The pioneering efforts applied to MEDLINE should, when applicable, be incorporated into the TEHIP program. The committee recommends that a two-step approach be implemented in making refinements to the TEHIP databases and emphasizes the need for evaluation components to be incorporated and then monitored throughout this process. The first step would entail improvements that could be made in the short term and while the user analysis is being conducted. The second step could be implemented over the long term and would be based on the results of the user analysis. Once the results of the user analysis are examined and it has been determined which of the TEHIP databases are most useful to health professionals, then a prioritization of activities (including outreach and training, access, and navigation) should be undertaken around those most useful databases.

The committee recommends that in the short term (during the time that the user analysis is being conducted) NLM continue its efforts to increase access to the TEHIP databases and to simplify navigation of the databases by coordinating the development of the TEHIP Experimental World Wide Web Interface with Internet Grateful Med, promoting online registration for database access, and exploring the possibilities of linking TEHIP's World Wide Web site with the Web sites of other health professional organizations and establishing pointers to the TEHIP databases from World Wide Web search engines.

The committee recommends that in the long term and on the basis of the priorities set as a result of the user analysis, NLM expand its efforts to facilitate access and navigation of the TEHIP databases by making full use of the navigational tools being developed within NLM and beyond. This includes implementation of more efficient and intuitive user interfaces, incorporation of the Unified Medical Language System (UMLS) knowledge sources and other expert systems that would enhance symptom-related and other natural language searches, incorporation of navigational tools and interfaces that would create a seamless interface between databases, and implementation of indicators of peer review into new search interfaces.

PROGRAM ISSUES

As the committee conducted this study, several programmatic issues came to the forefront. It is clear that NLM has taken a great deal of initiative in disseminating toxicology and environmental health information; however, there is a need for a stabilized funding base, an internal commitment to the TEHIP program and involvement of the TEHIP program in broader NLM research and development (R&D) efforts, an interdisciplinary advisory committee, and the development and implementation of a TEHIP-specific evaluation plan. The

TEHIP program continues to make a substantial contribution to the fields of toxicology and environmental health, and the beneficial use of toxicology and environmental health information resources by health professionals and other interested user communities can be considerably increased given the necessary resources and support.

The committee recommends that the TEHIP program be given the responsibilities and resources needed to strengthen its growth and development. This may be accomplished by providing a stable funding base, ensuring a leadership role for the TEHIP program and promoting the incorporation of the TEHIP program into broader technological developments at NLM, establishing an interdisciplinary advisory committee, and the implementation of an evaluation plan.

FUTURE DIRECTIONS

The committee believes that it is important to provide health professionals with the tools needed to retrieve toxicology and environmental health information. During the course of its deliberations, the committee determined that the optimal approach would be to provide health professionals with two different options for accessing toxicology and environmental health information. Primary access would be via an online directory or "road map" that would assist the user in identifying and connecting with the relevant TEHIP database or other (non-TEHIP) online information resource (e.g., Web sites of government agencies, directory information for environmental health organizations).[4] The second access option would be a toll-free telephone number or similar single-access information center that would link health professionals (person-to-person) with a specialist who could provide consultative services on environmental health issues and concerns.

This two-part approach would require cooperative efforts between NLM, other federal agencies, and private-sector organizations. NLM's expertise and current R&D efforts in medical informatics would make it the logical agency to take the leadership role for the online directory, although the assistance of other agencies and organizations would be needed to provide directory information. The development and implementation of the second option, the single-access information center, go beyond NLM's mandate, and this recommendation should be considered by multiple federal agencies and other private-sector or-

---

[4]The committee's vision of the online directory includes simple interactive screens that walk the user through various options for reaching the appropriate information resource through hypertext links, graphical interfaces, and search engines.

ganizations involved in environmental health. Public-private partnerships could play a key role in providing these information services. The collaboration of NLM with universities, industry, international resources, and local, state, and federal governments could be particularly productive given the numerous information resources in this field.

The implementation of a single-access information center would have budgetary and interagency ramifications that the committee did not have the mandate to explore. In its exploration of the TEHIP databases and its inquiries into the dissemination of toxicology and environmental health information—especially to health professionals—the committee realized the need for this type of information resource for toxicology and environmental health information. Poison control centers are examples of single-access points that are effectively meeting the information needs of health professionals and other user communities, and the committee believes that poison control centers would make excellent resources for providing toxicology and environmental health information to health professionals. However, the committee is particularly mindful of the budgetary considerations and of the problems that poison control centers would face if their mandate were expanded without the necessary financial resources for implementation. Although the provision of a single-access information center for toxicology and environmental health information is not within the purview of NLM, the issue should be explored because it is important to expanding the use of this information by health professionals.

> **The committee recommends that NLM, other relevant federal agencies, and private-sector organizations work cooperatively to provide health professionals and other interested user communities with the tools that they need to access toxicology and environmental health information. This would involve two different types of access points:**
>
> • **an online directory that would contain information on the full spectrum of information resources in toxicology and environmental health and that would direct the user to the appropriate online information resource; and**
> • **a single-access information center (e.g., regional poison control centers) that would connect the user with individuals with expertise in environmental health.**

1

Introduction

Health professionals need access to environmental health[1] and toxicology information for many reasons. Certainly, public awareness about human health risks from chemical and biologic agents in the environment has increased dramatically in recent years. Similarly, changing trends in health care and an emphasis on prevention, coupled with increasing computer literacy, all support the need for readily available information about the impacts of hazardous substances in the environment on individual and public health. Reports in the popular press and news media have highlighted the public's concern. For example, pesticides on foods; second-hand tobacco smoke; asbestos and lead paint in homes and public buildings; dioxin contamination; occupational exposures to gasoline and other chemicals; exposure to radon and benzene; and drinking water contaminated with biologic and chemical agents are just a few of the issues that may confront the American public.

Although the public relies heavily on federal and state regulatory agencies for protection from exposures to hazardous substances, they frequently look to health professionals for information on routes of exposure and the nature and extent of associated health consequences. However, most health professionals acquire only a minimal knowledge of toxicology during their education and training. As a result, their working knowledge of the adverse effects of chemicals on health and the conditions under which those effects might occur is often limited. Furthermore, with the many competing demands on health professionals' time, it is difficult, even for specialists, to keep apprised of rapidly evolving

---

[1]The committee's use of the term *environmental health* includes health issues regarding exposures to hazardous substances in the workplace, home, and community settings.

toxicology information. Thus, health professionals need ready access to toxicology and environmental health information resources to assist them with patient care. Policymakers, health advisors, researchers, health educators, and other involved communities (e.g., the general public) also need access to this information as they pursue their own inquiries.

Established in 1956 (Public Law 84-941), the National Library of Medicine (NLM) is charged with improving the nation's health by collecting and providing access to the world's biomedical literature. In 1967, NLM established a specialized information program in toxicology and environmental health known today as the Toxicology and Environmental Health Information Program (TEHIP).[2] Its mission is to provide selected core information resources and services, facilitate access to national and international information resources, and strengthen the information infrastructure of toxicology and environmental health (NLM, 1995). Currently, the TEHIP program encompasses 16 online databases that contain bibliographic and factual information on hazardous substances, including chemical properties, carcinogenicity, exposure levels, adverse health effects, emergency treatment protocols, and federal regulations. Additionally, the TEHIP program is responsible for training and outreach efforts to health professionals.

In 1995, at the request of NLM, the Institute of Medicine (IOM) formed the Committee on Toxicology and Environmental Health Information Resources for Health Professionals. The committee was charged with producing a consensus report examining the utility and accessibility[3] of NLM's TEHIP program for the work of health professionals and providing NLM with recommendations and strategies to improve use of the TEHIP databases. Additionally, the committee was asked to consider the current toxicology and environmental health information needs of health professionals and how those needs are currently being met. To fulfill this charge, the IOM selected for membership on its committee, individuals with expertise in a variety of disciplines, including medical and clinical toxicology, occupational and environmental health, primary care, library science and medical informatics, environmental science, health education, and emergency medicine.

The committee met three times during the course of the study and received extensive input from health professionals representing a range of disciplines and expertise. Input was received through several mechanisms. In conjunction with several health professional organizations, the committee developed and distributed a questionnaire that focused on the current use of computers and of online information resources in toxicology and environmental health (see Appendix B).

---

[2]The program was originally called the Toxicology Information Program (TIP) and was mandated to create automated toxicology data banks and provide toxicology information and data services.

[3]Factors examined in assessing the utility (usefulness) and accessibility (ease of use) of the databases include the subject content, search interface, and available access points.

INTRODUCTION

The committee received 247 responses to the questionnaire. Additionally, the committee sponsored a workshop during which thirty-four health professionals attended one of four focus group sessions that discussed information needs and information resources in toxicology and environmental health (see Appendix C). Workshop participants also attended a demonstration of the TEHIP databases presented by NLM staff. The committee received valuable input from the speakers who attended its meetings, including NLM staff members, occupational and environmental health specialists, and representatives from several government agencies involved in environmental health issues. In addition, the committee benefited from the work of the 1993 NLM Long Range Planning Panel on Toxicology and Environmental Health (NLM, 1993), which presented many of the issues and recommendations that the IOM committee considered and reaffirms throughout this report.

HEALTH PROFESSIONALS AND OTHER USER COMMUNITIES

Although toxicology and environmental health information is used in a wide range of occupations and professional endeavors, including nursing and clinical medicine, pharmaceutical development, chemical manufacturing, environmental engineering, and law, the committee was charged specifically with examining the toxicology and environmental health information needs of health professionals. This charge is in concert with the mission of NLM, which focuses on "the dissemination and exchange of scientific and other information important to the progress of medicine and to the public health" (NLM, 1986). The committee took a broad perspective that encompasses a number of groups with interests in environmental health (see also Chapter 4). The committee realizes that the health professional community does not have homogeneous information needs and there is a wide range of variation in access to online databases and other information resources. Even with these variations, however, the committee believed that it was worthwhile to discuss potential user communities to provide generalized insights into how NLM might better meet the toxicology and environmental health information needs of these groups. The following list of potential user communities is not meant to be definitive or exhaustive but rather was used by the committee for purposes of discussion:

- primary care professionals (e.g., physicians, nurses, nurse practitioners, and physician assistants) and pharmacists;
- specialists in occupational and environmental health (e.g., physicians, nurses, nurse practitioners, physician assistants, industrial hygienists, and safety officers);

- emergency medicine and poison control center personnel (e.g., emergency room health professionals, emergency medical technicians, clinical and medical toxicologists, and specialists in poison information);
- health science librarians and faculty at health professional schools (including medical, nursing, public health, pharmacy, and dental schools);
- environmental health researchers and scientists (including health physicists, epidemiologists, toxicologists, and forensic practitioners);
- patients, the general public, and community organizations (including local emergency planning committees, public librarians, educators, and advocacy and activist organizations); and
- health professionals in local public health departments or in state and federal agencies (e.g., policy advisors, health educators, and public clinic personnel).

Although the type, depth, and frequency of toxicology and environmental health information needed by each of these groups will differ among individuals within and across the groups, depending on job responsibilities, demographics, training, work or practice setting, time, access, and availability, these groupings provide a framework from which to explore information needs, current strategies for finding information, and potential use of the TEHIP databases.

This chapter provides the context for the report by discussing the ways in which both the public health impacts of hazardous substances and the changing trends in health care are reinforcing the need for authoritative and easily accessible information in the fields of toxicology and environmental health.

PUBLIC HEALTH IMPACTS OF HAZARDOUS SUBSTANCES

The proliferation of synthetic organic chemicals has created many hazards to human health and the environment. Furthermore, many of these compounds are persistent in the environment, accumulate in human tissue, and are inadequately tested for both their ecological and human health effects. Pathways of human exposure to chemicals in the environment are diverse. Chemicals in air, food, water, and soil can be inhaled, ingested, or absorbed in any number of settings. Such exposures can occur at home, work, and school, as well as in vehicles, public buildings, and outdoor community environments. Vulnerable populations, such as children, the elderly, the chronically ill, minorities, and the poor may be at increased risk of harm related to environmental contamination because of biologic and demographic factors, including where they live.[4] The

---

[4]The Centers for Disease Control and Prevention (CDC) estimates that 3 million to 4 million children in the United States (many living in older houses with lead paint) have blood lead levels above 10 µg/dl—a level that may cause neurologic effects, including

Environmental Protection Agency (EPA) estimates that one in four Americans (including 10 million children under the age of 12) lives within 4 miles of a toxic waste dump (EPA, 1996).

Determining the human health effects of environmental contamination is a complex undertaking. People are exposed to many chemicals simultaneously, and little is known about the health consequences of multiple chemical exposures (INFORM, 1995; IOM, 1995). Most human studies focus on either occupational groups or clinical cases involving acute poisonings, both of which generally involve higher exposure levels than those experienced by the general population. There is limited information about the health effects associated with low-level chemical exposure over extended periods of time. The technical difficulties of measuring exposure, determining the actual dose to the target organs, and extrapolating data from animal studies to human populations further confound the ability to assess the effects of hazardous substances on human health.

Despite these difficulties, the roles of certain hazardous substances in the development of human disease are well-known (Lybarger et al., 1993). Environmental chemical exposures can affect all organ systems. They can cause or contribute to the development of a variety of human illness, including cancer, asthma and other respiratory diseases, reproductive disorders, neurological and immune system impairments, and skin disease, as well as cardiovascular, renal, hepatic, and psychological disorders (Rom, 1992).

The proliferation of the manufacture, transport, use, and disposal of chemicals coupled with the potential for human exposures makes it essential that health professionals have easy access to resources and information to assist them in the prevention, diagnosis, and treatment of disorders possibly stemming from exposures to hazardous substances in the environment.

CHANGING TRENDS IN HEALTH PRACTICE

In this time of major changes in health care in the United States, many individuals are becoming better informed health care consumers and more involved in their own health care. In assuming a more involved role in prevention and health care, individuals often seek out information that can improve or maintain their health, as well as information about preventing, managing, and treating disease and illness. When patients are unable to independently find or fully understand information related to their environmental health concerns, they may turn to their primary care physician or other trusted health professional for accurate information and answers (IOM, 1988). Since disease prevention is an area of growing interest and activity, it is important that health professionals

---

deficits in attention and IQ scores (ATSDR, 1988; CDC, 1991; Needleman et al., 1979, 1990).

be aware of prevention options and knowledgeable enough to counsel patients with possible environmental exposures. Examples of primary prevention efforts in environmental health include reducing radon exposure to prevent lung cancer, installing carbon monoxide detectors, and counseling people with asthma on choice of workplace and recreation. Additionally, right-to-know laws in the workplace and community as well as consumer product labeling place information demands on health professionals as patients, workers, and community groups ask questions about exposure risk and potential adverse health effects.

The rapid growth of biomedical knowledge and the resulting increase in the number of scientific journals that have inundated health professionals are other factors affecting the need for accessible toxicology and environmental health information (Deering and Harris, 1996). Approximately 300,000 new references are added to the MEDLINE database annually (Hersh, 1996). This "information explosion" has resulted in the greater use of computers to store and rapidly retrieve information (Hersh, 1996; Huth, 1989). Additionally, the increasing emphasis on cost-effectiveness in health care has made health professionals aware of the marginal utility of their information-seeking actions and the need to use the most cost-effective mechanism for locating information (Greenes and Shortliffe, 1990). Having authoritative information rapidly available can have a significant impact on health care treatment, particularly in emergency situations where time is the critical factor. Poison control centers are a prime example of cost-effective information resources. As computer applications become the standard in health care, health professionals are becoming increasingly adept at using computers. The use of an evidence-based approach[5] to health care may increase the emphasis on the analysis of medical literature in clinical decision-making and may have an impact on the use of online bibliographic databases. Thus, the committee believes that both current trends in health care and heightened concerns about environmental health issues are increasing the need for online information resources in toxicology and environmental health.

ORGANIZATION OF THE REPORT

Chapter 1 has provided a brief overview of the background of this report and the context in which the committee discussed the need for authoritative and accessible information resources in toxicology and environmental health. Chapters 2 and 3 focus on the broad spectrum of toxicology and environmental health information resources currently available. Specifically, Chapter 2 provides an overview of NLM's TEHIP program and briefly describes the program's 16 databases, and Chapter 3 focuses on other toxicology and environ-

---

[5]This approach involves the retrieval of relevant medical literature, critical evaluation of the validity of the studies based on methodological rigor, and basing patient care decisions on the weight of the evidence (Evidence-Based Medicine Working Group, 1992).

mental health information resources and considers NLM's role in the context of this spectrum of information. Chapter 4 identifies the various groups of health professionals that have an interest in toxicology and environmental health information and examines the methods by which health professionals locate information. Chapter 5 explores NLM's current training and outreach efforts and presents the committee's ideas on future directions to increase health professionals' awareness of the available information resources. Chapter 6 considers the issues involved in accessing and navigating the TEHIP databases and presents short- and long-term recommendations for addressing the barriers faced by database searchers. The final chapter, Chapter 7, discusses programmatic issues that will affect the future growth and development of the TEHIP program and presents the committee's recommendations for tools that will assist health professionals in retrieving the toxicology and environmental health information that they require.

During the course of this study, the committee discussed at length how best to converge NLM's strengths in organizing and disseminating biomedical information with the disparate nature and location of toxicology and environmental health information and the wide variety of health professionals' information needs. The committee concluded that it is critical for NLM to exert its leadership role in biomedical information in the fields of toxicology and environmental health. As will be discussed throughout the report, the committee emphasizes the need for NLM to take a library-focused approach encompassing the full spectrum of toxicology and environmental health information, in addition to the 16 TEHIP databases.

This report presents recommendations to NLM regarding the TEHIP program. Additionally, the committee considered the larger issue of organizing the vast quantity of databases and other information resources on toxicology and environmental health and makes recommendations on establishing central access points for information in this field. As a result, the intended audience for this report extends beyond NLM staff to include government and private sector information providers as well as policymakers, librarians, and health professionals involved in providing and using toxicology and environmental health information.

REFERENCES

ATSDR (Agency for Toxic Substances and Disease Registry). 1988. *The Nature and Extent of Lead Poisoning in Children in the U.S.: A Report to Congress.* Atlanta, GA: ATSDR.

CDC (Centers for Disease Control and Prevention). 1991. *Preventing Lead Poisoning in Young Children.* Atlanta, GA: CDC.

Deering MJ, Harris J. 1996. Consumer health information demand and delivery: Implications for libraries. *Bulletin of the Medical Library Association* 84(2):209–216.

EPA (Environmental Protection Agency). 1996. *A National Agenda to Protect Children's Health from Environmental Threats*. Washington, DC: EPA.

Evidence-Based Medicine Working Group. 1992. Evidence-based medicine: A new approach to teaching the practice of medicine. *Journal of the American Medical Association* 268(17):2420–2425.

Greenes RA, Shortliffe EH. 1990. Medical informatics: An emerging academic discipline and institutional priority. *Journal of the American Medical Association* 263(8): 1114–1120.

Hersh WR. 1996. *Information Retrieval: A Health Care Perspective*. New York: Springer-Verlag.

Huth EJ. 1989. The information explosion. *Bulletin of the New York Academy of Sciences* 65(6):662–672.

INFORM. 1995. *Toxics Watch, 1995*. New York: INFORM, Inc.

IOM (Institute of Medicine). 1988. *Role of the Primary Care Physician in Occupational and Environmental Medicine*. Washington, DC: National Academy Press.

IOM. 1995. *Nursing, Health, and the Environment: Strengthening the Relationship to Improve the Public's Health*. Washington, DC: National Academy Press.

Lybarger JA, Spengler RF, DeRosa CT. 1993. *Priority Health Concerns: An Integrated Strategy to Evaluate the Relationship Between Illness and Exposure to Hazardous Substances*. Atlanta, GA: ATSDR.

Needleman HL, Gunnoe C, Leviton A, Reed R, Peresie H, Maher C, Barrett P. 1979. Deficits in psychologic and classroom performance of children with elevated dentine lead levels. *New England Journal of Medicine* 300(13):689–695.

Needleman HL, Schell A, Bellinger D, Leviton A, Allred EN. 1990. The long-term effects of exposure to low doses of lead in childhood: An 11-year follow-up report. *New England Journal of Medicine* 322(2):83–88.

NLM (National Library of Medicine). 1986. *Locating and Gaining Access to Medical and Scientific Literature. Long Range Plan, Report of Panel 2*. Bethesda, MD: NLM.

NLM. 1993. *Improving Toxicology and Environmental Health Information Services*. Report of the Board of Regents Long Range Planning Panel on Toxicology and Environmental Health. NIH Publication No. 94-3486. Bethesda, MD: NLM.

NLM. 1995. *National Library of Medicine Fact Sheet: Toxicology and Environmental Health Information Program*. Rockville, MD: NLM.

Rom WN, ed. 1992. *Environmental and Occupational Medicine*, 2nd ed. Boston: Little, Brown, and Company.

2

The National Library of Medicine's Toxicology and Environmental Health Information Program

In 1966, at a time of increased concern over the potential health effects of chemicals, the President's Science Advisory Committee examined the state of information in the growing science of toxicology and concluded that "there exists an urgent need for a much more coordinated and more complete computer based file of toxicological information than any currently available and, further, that access to this file must be more generally available to all those legimately needing such information" (PSAC, 1966). That recommendation led to the establishment of the Toxicology Information Program (TIP) at the National Library of Medicine (NLM), which was retitled in 1994 as the Toxicology and Environmental Health Information Program (TEHIP) to more accurately reflect the scope of the program. The TEHIP program currently encompasses 16 databases offering a wide range of toxicology and environmental health information of importance to health professionals, the general public, scientists, and policymakers.

This chapter provides a brief overview of NLM and the Division of Specialized Information Services (SIS), which oversees the TEHIP program. To inform readers unfamiliar with the breadth and depth of the TEHIP program's information resources, the main focus of this chapter is a description of the TEHIP program and of each of the TEHIP databases. The chapters that follow provide an assessment of the TEHIP program and address the toxicology and environmental health information needs of health professionals. For definitions of the acronyms refer to the glossary included at the end of the report.

NATIONAL LIBRARY OF MEDICINE

Begun in the early 1800s as a part of the Office of the Surgeon General of the Army, NLM has expanded its mission and scope to become one of the country's three national libraries[1] and the primary collector of medical information (Miles, 1982). NLM's collection exceeds 5 million books, journals, and audiovisuals on health and medicine. The library also provides access to more than 40 online databases through its Medical Literature Analysis and Retrieval System (MEDLARS) (NLM, 1995). One of NLM's most well-known achievements is the bibliographic database MEDLINE, which is internationally respected as a source of citations and abstracts to the world's medical literature. Approximately 300,000 citations are added to the MEDLINE database annually.

The Lister Hill National Center for Biomedical Communications and the National Center for Biotechnology Information are the research and development arms of NLM. Through these centers, NLM conducts a range of intramural and extramural research to explore new technologies in the fields of medicine, library science, computer science, and informatics.

The National Library of Medicine Act of 1956 (Public Law 84-941) broadly mandated that NLM collect and organize health sciences information "to assist the advancement of medical and related sciences, and to aid the dissemination and exchange of scientific and other information important to the progress of medicine and to public health" (NLM, 1986). The library works with the 4,500 health science libraries of the National Network of Libraries of Medicine (NN/LM) and the eight Regional Medical Libraries (covering all geographic regions of the United States) to provide health professionals and other interested individuals with access to biomedical information. More than 3 million interlibrary loan requests are filled annually by NN/LM.

Organization and Funding

NLM is organized into six divisions (Figure 2.1), the largest of which is Library Operations, responsible for fundamental library services including literature collection, indexing, and cataloging. In fiscal year (FY) 1995, NLM employed 586 full-time equivalent (FTE) personnel.

---

[1]The Library of Congress and the National Agricultural Library are the other national libraries.

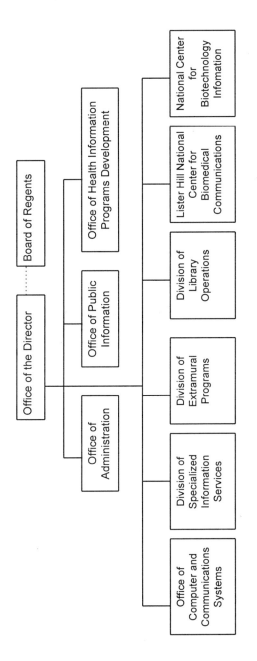

FIGURE 2.1 National Library of Medicine organizational chart. SOURCE: NLM, 1995.

The majority of NLM's funding comes from its federal budget appropriation, which in FY 1995 was $128 million. Additionally, NLM receives reimbursements for projects with other agencies ($12.9 million in FY 1995). Although user fees are charged for searching the NLM database system,[2] these fees do not supplement the NLM budget. Appropriated funds are used to build and maintain the MEDLARS databases, and user fees cover only the added costs associated with accessing this information (e.g., telecommunications charges) (NLM, 1996b,c).

MEDLARS

In the early 1960s, NLM applied emerging computer technologies to establish MEDLARS, which was used to produce bibliographic publications including *Index Medicus* and to conduct individual literature searches for health professionals (Miles, 1982). Since then MEDLARS has grown to encompass more than 40 bibliographic and factual databases, of which the most well-known and most often searched is the bibliographic database MEDLINE. In 1995, more than 7.3 million searches were performed on the MEDLARS databases (NLM, 1996a). There are a number of specialized databases on MEDLARS including those pertaining to HIV/AIDS (e.g., AIDSLINE and AIDSDRUGS), bioethics (BIOETHICSLINE), and the history of medicine (HISTLINE). Additionally, the toxicology and environmental health databases discussed in this report are part of MEDLARS. Table 2.1 provides a timeline overview of some of the major events and changes occurring within the past 40 years in computer technology, in the emergence and response to environmental health issues, and in the development of NLM's online databases.

MEDLARS databases reside on two separate computer subsystems. NLM's original retrieval system, ELHILL, was designed for bibliographic databases. As a result, in the early 1980s when the records of the TEHIP factual databases became too large for the ELHILL system, it was determined that a new system was needed. TOXNET was developed in 1985 and is the system of networked microcomputers used for database file building, updates, and online searching for most of the TEHIP databases (Van Camp, 1989). A gateway links the ELHILL and TOXNET systems, making the databases available for searching by all NLM users.

---

[2]Fees are also charged for leasing NLM databases.

TABLE 2.1 Timeline of Events and Changes in Computer Technology and Environmental Health

Computer Technology and the MEDLARS Databases		Environmental Health and the TEHIP Program	
NLM established	1956		
Batch processing; computers adopted for data processing by corporations	1960s		
Punched card batches used to produce *Cumulated Index Medicus*	1961	1961	Society of Toxicology founded
		1962	Publication of Rachel Carson's book *Silent Spring*
MEDLARS introduced; requested searches were batch processed	1964		
		1965	IARC established by the World Health Organization
		1966	Publication of the President's Science Advisory Committee's report *Handling of Toxicological Information*
		1967	NLM's TIP established
Time-sharing (sharing of computer services by multiple users introduced)	1970s	1970	Occupational Safety and Health Act established NIOSH and OSHA
			Clean Air Act enacted
Microprocessor developed; enabled the development of the personal computer	1971		
MEDLINE introduced online			
ARPAnet, the precursor to the Internet, becomes operational	1972	1972	TOXLINE developed by NLM
			Clean Water Act enacted
MEDLINE tapes leased to commercial vendors	mid-1970s		

continues

TABLE 2.1 Continued

Computer Technology and the MEDLARS Databases			Environmental Health and the TEHIP Program
		1976	Toxic Substances Control Act (TSCA) enacted
		1978	Toxicology Data Bank developed by NLM
			National Toxicology Program established
Desktop computers	1980s	1980	Comprehensive Environmental Response, Compensation, and Liability Act (CERCLA) enacted; ATSDR established
		1983	OSHA's Hazard Communication Standard, which covered employees in the manufacturing sector of industry, is enacted
		1985	TOXNET developed
Grateful Med developed by NLM	1986	1986	Superfund Amendments and Reauthorization Act (SARA) enacted
		1987	OSHA's Hazard Communication Standard expanded to include employees in all industries
		1988	IOM report *Role of the Primary Care Physician in Occupational and Environmental Medicine* published
		1989	Toxic Chemical Release Inventory (TRI) online database

continues

TABLE 2.1 Continued

Computer Technology and the MEDLARS Databases			Environmental Health and the TEHIP Program
University of Minnesota introduces Gopher which rapidly increases access to the Internet	1990		
NSCA releases first version of the Web browser, Mosaic	1992		
		1993	NLM Long Rang Planning Panel report *Improving Toxicology and Environmental Health Information Services* is published
Number of Internet hosts reaches 6 million	1995	1995	IOM reports *Environmental Medicine* and *Nursing, Health, and the Environment* published
Internet Grateful Med developed by NLM	1996	1996	National Occupational Research Agenda established by NIOSH

NOTE: ATSDR=Agency for Toxic Substances and Disease Registry; EPA=Environmental Protection Agency; IARC=International Agency for Research on Cancer; NIOSH=National Institute for Occupational Safety and Health; NLM=National Library of Medicine; NSCA=National Center for Supercomputing Applications; OSHA=Occupational Safety and Health Administration; TIP=Toxicology Information Program.

SOURCES: Brooks et al. (1995), Campbell-Kelly and Aspray (1996), Kissman and Wexler (1985), Miles (1982), Netscape (1996), Rom (1992), Tesler (1991), Zenz et al. (1994).

DIVISION OF SPECIALIZED INFORMATION SERVICES

Since the 1960s, NLM has had a commitment to the collection and dissemination of toxicology and environmental health information. The Specialized Information Services (SIS) Division of NLM is responsible for the TEHIP program in addition to its responsibilities for the AIDS-related databases (AIDSLINE, AIDSTRIALS, and AIDSDRUGS) and other designated activities, including outreach programs.

There are 34 FTE personnel working in SIS, making it the third smallest of the NLM divisions (NLM, 1995). SIS is organized into two branches, Biomedical Information Services and Biomedical Files Implementation. Personnel in both branches work on the TEHIP program, and many have responsibilities beyond TEHIP.

TOXICOLOGY AND ENVIRONMENTAL HEALTH INFORMATION PROGRAM

Mission and History

The 1966 President's Science Advisory Committee report *Handling of Toxicological Information* provided the impetus for the development of the Toxicology Information Program (TIP) at NLM (PSAC, 1966). This program began in 1967 with the anticipation of annual funding of several million dollars and staffing of 40 FTEs. However, funding and staffing never reached these anticipated levels. Instead, over the next several years the budgets for TIP were approximately $1 million annually and staffing levels remained at 10 to 18 people (Miles, 1982; NLM, 1993).

TIP undertook a variety of projects, including answering reference queries from the biomedical community. Because staff levels at TIP remained lower than anticipated, NLM initiated an interagency agreement in 1972 with the Atomic Energy Commission to establish a Toxicology Information Response Center at Oak Ridge National Laboratory (ORNL) to handle the reference inquiries (NLM, 1993).[3] Additionally, TIP produced numerous publications including the Pesticides Abstracts series—a joint project with the Environmental Protection Agency (EPA).

The primary focus of TIP was the development of online databases, beginning with the TOXLINE database in 1972 (Kissman and Wexler, 1985). TOXLINE was designed as a comprehensive bibliographic resource for scientific literature on toxicology. As described below, TIP developed other databases to meet the needs of users searching for information on chemicals, including the dictionary files CHEMLINE and ChemID and the Hazardous Substances Data Bank (HSDB), an encyclopedic factual database originally developed as the Toxicology Data Bank.

Throughout the program's 29-year history, other databases have been added, many of which originate in other federal agencies, including EPA, the National Cancer Institute (NCI), and the National Institute for Occupational Safety and Health (NIOSH). These databases have been added to the NLM system primarily in response to legislative mandates or because of the agency's interest in making its databases accessible to a wider audience. For example, the

---

[3]This center continues to operate as an independent function of ORNL, with referrals made to it by NLM staff but without NLM funding.

Superfund Amendments and Reauthorization Act of 1986 (SARA) mandated the collection and electronic dissemination of information on the annual release of chemicals into the environment by industrial facilities. As a result, EPA developed the Toxic Chemical Release Inventory database (TRI) and has disseminated the TRI database through NLM since 1989 (beginning with data submitted for TRI87, the first of these annual compilations). The Comprehensive Environmental Response, Compensation, and Liability Act of 1980 (CERCLA) mandated the establishment of the Agency for Toxic Substances and Disease Registry (ATSDR) and specified that ATSDR maintain an inventory of the health effects of toxic substances. This legislation led to collaborative efforts between NLM and ATSDR in the expansion of what had been the Toxicology Data Bank to become the HSDB. Thus, the evolution of NLM's TEHIP program has been the result of both internal NLM commitments to developing toxicology and environmental health information resources and the interests of other federal agencies in fulfilling their missions and legislative mandates.

In early 1994, following the recommendations of the NLM Long Range Planning Panel (NLM, 1993), the program's focus on environmental health information was made more explicit by renaming the program as the Toxicology and Environmental Health Information Program (TEHIP). The mission of the TEHIP program is to:

- provide selected core toxicology and environmental health information resources and services,
- facilitate access to national and international toxicology and environmental health information resources, and
- strengthen the information infrastructure of toxicology and environmental health (NLM, 1996d).

The TEHIP program encompasses the 16 databases described below. It also offers other services and programs and performs extensive outreach efforts (see Chapter 5). Other aspects of the TEHIP program are addressed throughout the report.

Funding

As with NLM as a whole, the TEHIP program receives funds from two sources: those appropriated to NLM by the U. S. Congress and reimbursements from other agencies of the federal government. When the figures are adjusted for inflation, it can be seen that the budgeted funding for the TEHIP program has remained relatively constant over the past 29 years (in FY 1994, the TEHIP program's appropriated budget in current dollars was approximately $7.4 million [NLM, 1995]; Figure 2.2). However, TEHIP program reimbursements

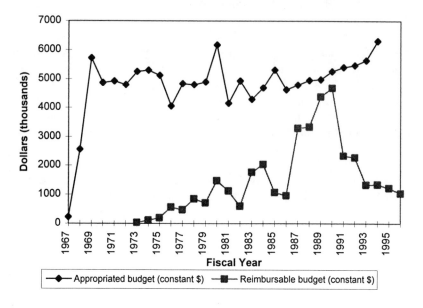

FIGURE 2.2 TEHIP program budget (constant dollars).

from other agencies have fluctuated. Reimbursement funding is the result of collaborative projects with other federal agencies. Recently, the TEHIP program's financial reimbursements from other agencies have dropped significantly.[4] In FY 1992, the TEHIP program's total reimbursable budget from other agencies was $2.45 million, whereas in FY 1993 the reimbursable budget dropped by approximately 50 percent to $1.27 million. Since 1993, the reimbursable budget has remained relatively level (the FY 1995 reimbursable budget was $1.23 million).

---

[4]The reimbursable funds from the National Technical Information Service (NTIS) and ATSDR had the largest percentage decreases in the 1990s. Decreases in NTIS funds resulted from changes in NLM policy on how NTIS funds could be used for NLM projects. During the same time period, ATSDR developed its own in-house information resources (e.g., the HazDat database) and had less need to fulfill its mission by supporting information resources managed by other agencies such as NLM.

TEHIP DATABASES

The TEHIP program has 16 online databases (Table 2.2). These databases originate in several different federal agencies, a fact that has complicated attempts to standardize the databases and improve access. One of the features of the TEHIP databases unfamiliar to many users is the inclusion of both bibliographic and factual databases.[5]

Bibliographic databases are fairly standard in format and are organized to have one record per article citation (Figure 2.3). Each record includes the reference information needed to identify a journal article or other document (e.g., author, title, source, volume, and page numbers) and an abstract if one is available. To enhance the precision of online searching, Medical Subject Headings (MeSH) indexing terms are often included in NLM bibliographic records (Lowe and Barnett, 1994). Many health professionals are familiar with MeSH from their searches of MEDLINE. MeSH is a controlled vocabulary that is used to describe the subject content of the journal article.

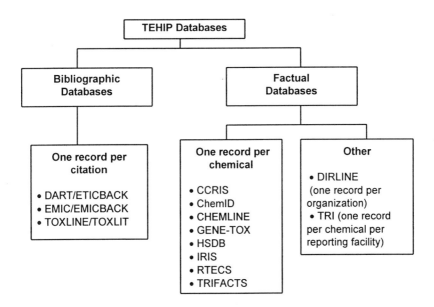

FIGURE 2.3 Organization of the TEHIP databases.

---

[5]Bibliographic databases contain citations (and often abstracts) to the scientific literature; searchers must go to another source to obtain the full text of the cited article (Siegel et al., 1990). Factual databases contain actual data (e.g., a chemical's physical and toxic properties) excerpted for inclusion in the database and usually provide a citation to the original source of the information.

The common feature of factual databases is their presentation of factual data (Table 2.3); however, this information can vary widely in content and form and may include original data from scientific studies, summary statements, dictionary-type information (e.g., names and synonyms), or directory-type information (e.g., addresses and phone numbers). Most of the TEHIP factual databases are organized by chemical record (i.e., each record in that database contains the information on one specific chemical). Exceptions to this are the TRI and DIRLINE databases, whose unique features are described below.

The following sections provide brief descriptions of the TEHIP databases—their subject content, organization, and review and update procedures. Further detailed information on each database, including descriptions of each of the searchable fields, is available in the NLM publication *Reference Materials for the Toxicology Information Program Online Services* (NLM, 1994). Technical information can be obtained from the responsible federal agency.

TEHIP FACTUAL DATABASES

Ten of the 16 TEHIP databases are factual databases and range from presenting original experimental data to summaries reflecting scientific judgments of the toxicity and risks associated with exposure to chemical substances.

ChemID and CHEMLINE

Information on chemicals can be difficult to locate because of the numerous synonyms and trade names by which chemicals are known and the complex nomenclature used to name chemicals, including the use of numerals and Greek letters. The Chemical Abstracts Service (CAS) Registry Number (RN) is a unique identifying number assigned by CAS to each chemical and is a searchable field in all of the TEHIP databases except DIRLINE. By using the Registry Number, the searcher ensures that the information pertains to the specified chemical, thereby relieving the searcher from knowing all the synonyms or the precise nomenclature.

The ChemID and CHEMLINE databases, both developed and maintained by NLM (CHEMLINE with CAS assistance), are dictionary-type files that provide the searcher with the CAS Registry Number or other information needed to hone a search on a specific chemical. Once that information is obtained the searcher can comprehensively search other TEHIP databases.

TABLE 2.2 TEHIP Databases[a]

Database	Sponsoring Agencies	Factual or Bibliographic	Number of Records	Subject Content	Review and Update
CCRIS	NCI	Factual	>7,100	Results of carcinogenicity, mutagenicity, tumor production, tumor inhibition studies	Studies selected by experts in respective fields, updated monthly
ChemID	NLM	Factual	>294,000	Chemical dictionary, synonyms, CAS Registry Number, molecular formula, SUPERLIST	Information from nonproprietary sources chosen by NLM, updated quarterly
CHEMLINE	NLM	Factual	>1.4 million	Chemical dictionary, synonyms, CAS Registry Number, molecular formula	Information from Chemical Abstracts and other sources, updated bimonthly
DART	NLM, NIEHS, FDA, EPA	Bibliographic	>29,000	Literature on teratology and many aspects of reproductive toxicology	Citations from MEDLINE and other sources, updated monthly
DIRLINE	NLM	Factual	>17,000	Directory of health information resources including organizations, software, and databases	Eleven sources of information, updated quarterly
EMIC	NLM, EPA, NIEHS	Bibliographic	>14,000	Literature since 1991 published on substances tested for genotoxic activity	Information from ORNL, updated monthly
EMICBACK	NLM, EPA, NIEHS	Bibliographic	>75,000	Pre-1950 through 1991 literature on substances tested for genotoxic activity	Information from ORNL, closed file

continues

TABLE 2.2 Continued

Database	Sponsoring Agencies	Factual or Bibliographic	Number of Records	Subject Content	Review and Update
ETICBACK	NLM	Bibliographic	>49,000	Teratology literature from 1950–1989, continued by DART	Information from ORNL, closed file
GENE-TOX	EPA	Factual	>2,900	Results from expert review of scientific literature on chemicals tested for mutagenicity	Reviewed by panels of external scientists, updates subject to EPA revisions
HSDB	NLM (previously ATSDR)	Factual	>4,500	Peer-reviewed summaries of the toxicology of potentially hazardous substances	Peer-reviewed by panel of external reviewers
IRIS	EPA	Factual	>660	EPA health risk and regulatory information, includes carcinogenic and noncarcinogenic risk assessment data	Reviewed by two working groups of EPA scientists, updated monthly
RTECS	NIOSH	Factual	>133,000	Toxic effects including skin and eye irritation, carcinogenicity, mutagenicity, and reproductive consequences	Information selected by NIOSH staff, updated quarterly

TOXLINE	NLM	Bibliographic	>2,000,000	Bibliographic information from 18 sources covering the toxicology literature	18 sources of information, updated monthly
TOXLIT	NLM	Bibliographic	>2,000,000	Bibliographic information from royalty sources covering toxicology literature	Information primarily from Chemical Abstracts, updated monthly
TRI	EPA	Factual	>75,000 (TRI94)	Annual estimated releases of toxic chemicals to the environment	Information from industrial facilities, annual updates
TRIFACTS	EPA	Factual	>300	Summarized information on the health effects, ecological effects, safety, and handling of TRI chemicals	Adapted from New Jersey Fact Sheets, update subject to EPA revisions

NOTE: CAS=Chemical Abstracts Services; FDA=Food and Drug Administration; and NIEHS=National Institute of Environmental Health Sciences.

[a] Abbreviations are defined in the text.

Organization and Content

The ChemID and CHEMLINE databases are organized by chemical record. Each record consists of up to 18 fields of data, including the CAS Registry Number, synonyms (e.g., the aspartame record lists 22 synonyms for the chemical including its trade name, Nutrasweet), molecular formula, MeSH headings, and number of rings and ring size. The classification code field lists the general categories (e.g., insecticide, vasodilator, or steroid) to which a chemical belongs. Another useful field found in both databases is the locator field, which lists all of the NLM databases[6] (including MEDLINE) that contain information (either factual or bibliographic) on the specific chemical (Box 2.1). Thus, a searcher could verify that an exhaustive search was done by checking that all databases with information on the chemical had been searched.

Chemicals are selected for inclusion in ChemID or CHEMLINE when they are listed in any one or more of the other NLM databases (including MEDLINE) or in the EPA Toxic Substances Control Act (TSCA) Inventory of Chemical Substances. Additionally, CHEMLINE contains records for chemicals on the European Inventory of Existing Commercial Chemical Substances.

ChemID has records on more than 294,000 chemical compounds. One of the unique and useful features of ChemID is the SUPERLIST field, which specifies the scientific and regulatory lists on which the chemical appears. Information is currently available for 19 listings including the EPA Pesticide List, the Occupational Safety and Health Administration (OSHA) Toxic and Hazardous Substances List, and the National Toxicology Program (NTP) Carcinogen List. ChemID specifies the particular spelling of the chemical name or the synonym used by each list and includes information on each list including a list description, contact information for the list producer, and references.

CHEMLINE has records on more than 1.4 million chemical substances and was the first of the two chemical dictionary databases to be included in the NLM online system. The data in CHEMLINE are primarily supplied by CAS with NLM augmenting the file with nonproprietary data. The royalty charges for the CAS data result in significantly higher connect fee charges for CHEMLINE than for ChemID (see Chapter 6). In 1989, concerned that the royalty charges would discourage access to CHEMLINE, NLM developed ChemID from non-royalty sources and provides it to users at regular MEDLARS rates.

---

[6]The locator field also lists two databases that are not on the NLM system: the TSCA Inventory (TSCAINV) and the European Inventory of Existing Commercial Chemical Substances (EINECS).

TABLE 2.3 Types of Information Available in the TEHIP Databases

Databases	Factual			Bibliographic
	Original/ Experimental Data	Summary Statements/ Data Excerpts	Dictionary or Directory Information	Bibliographic Citations and Abstracts
CCRIS		x		
ChemID			x	
CHEMLINE			x	
DART/ETICBACK				x
DIRLINE			x	
EMIC/EMICBACK				x
GENE-TOX	x	x		
HSDB		x		
IRIS	x	x		
RTECS		x		
TOXLINE/TOXLIT				x
TRI	x			
TRIFACTS		x		

> **BOX 2.1**
> **Locator Field in the ChemID Database**
>
> NM   Aspartame
> RN   22839-47-0
> LO   AIDSLINE; CANCERLIT; CCRIS; DART; EINECS; EMICBACK;
>   ETICBACK; HSDB; MEDLINE; MED75; MED80; MED85; MED90;
>   MESH; RTECS; TOXLINE; TOXLINE65; TOXLIT; TOXLIT65;
>   TSCAINV; SUPERLIST
>
> NOTE: NM = name; RN = Registry Number; and LO = locator field. The acronyms for the databases are listed in the glossary.

A computer-based tutorial, CHEMLEARN, has been developed by SIS to instruct librarians, information specialists, and scientists on effective searching strategies for the ChemID and CHEMLINE databases.

Hazardous Substances Data Bank

In 1978, NLM began development of the Toxicology Data Bank to provide an online source for evaluated toxicology data (Kissman and Wexler, 1985). This database has evolved into HSDB, a peer-reviewed database with a wide range of information on more than 4,500 chemicals. Funding for HSDB has been provided, in part, by ATSDR.

The user has three options for searching HSDB: direct searching (i.e., using the command language), menu searching, or using Grateful Med, a front-end software package developed by NLM. Direct searching requires the user to have knowledge of specific field names and the complex command language, whereas menu and Grateful Med searching provide useful entry points for infrequent users of the database.

Organization and Content

HSDB is organized by chemical record, and each record may contain more than 150 fields of information. The scope of this database broadly encompasses toxicology information, and HSDB records are often quite extensive. Standard texts, monographs, and other tertiary sources are the major sources for HSDB information, which are augmented by primary literature and with information from other databases (e.g., POISINDEX®[7]). NLM's decision to use authoritative

---

[7]POISINDEX® is a commercial database produced by Micromedex, Inc.

tertiary sources was based on the fact that these sources had already undergone a selection and evaluation process prior to being published (Kissman and Wexler, 1985). The Source Evaluation Team at NLM selects the HSDB sources on the basis of a number of criteria including peer review, quality, and conciseness (NLM, 1994).

Eleven major categories of information are available in each record (Box 2.2), each of which has numerous fields of information. As with any TOXNET database the searcher can retrieve the entire record for a chemical, look at all the information contained in a major category (e.g., toxicity or biomedical effects), or retrieve just individual fields of information (e.g., emergency medical treatment).

> **BOX 2.2**
> **Major Categories of HSDB Data**
>
> - Substance Identification
> - Manufacturing/Use Information
> - Chemical and Physical Properties
> - Safety and Handling
> - Toxicity/Biomedical Effects
> - Pharmacology
> - Environmental Fate/Exposure Potential
> - Exposure Standards and Regulations
> - Monitoring and Analysis Methods
> - Additional References
> - Express Data

The substance identification section includes the CAS Registry Number, molecular formula, and synonyms. HSDB includes extensive information on manufacturing and other uses for the chemical, including the most current U.S. production figures for that chemical. Information on chemical and physical properties includes 19 fields of information such as boiling point, molecular weight, spectral properties, and viscosity.

Extensive information on safety and handling that may be useful for emergency response teams is available. Information in this category includes the U.S. Department of Transportation (DoT) Emergency Guidelines, information on respirators and other protective equipment and clothing, clean-up and disposal methods, and facts on storage conditions.

HSDB includes a section on toxicity and biomedical effects that contains available human and animal data. NLM has a reciprocal agreement with Micromedex, Inc., in which information from the Micromedex database POIS-INDEX® is included in HSDB (e.g., information on emergency medical treatment including clinical effects, laboratory tests, treatment overview, and range of toxicity). In return, HSDB is made available as a subfile of Micromedex's

TOMES Plus® database (Toxicology, Occupational Medicine, and Environmental Series). This section of HSDB also draws from other databases and reports and has separate fields for data from the NTP Reports, the International Agency for Research on Cancer (IARC) Summary and Evaluation, and the TSCA Test Submissions (industry reports submitted to EPA).

Extended information is also available on the chemical's pharmacokinetics, pharmacology (if it is a drug), environmental fate and exposure potential (including food survey values, probable routes of human exposure, and human body burden information), monitoring and analysis methods (e.g., analytic and clinical laboratory methods), additional references, and the new information that has been added to the HSDB record (included in the express data field). The major federal exposure standards and regulations on the chemical are a part of the HSDB record and include the OSHA Standards, the NIOSH Recommendations, the CERCLA Reportable Quantities, and the Water Standards.

Review and Updates

HSDB is peer-reviewed; HSDB data initially undergo review by the NLM Quality Control group, a function provided to NLM under contract, in which checks are made for missing data, errors, lack of clarity, and so forth. Data that have passed this level of review are labeled as quality-control reviewed. The more extensive peer review process is conducted by the Scientific Review Panel, a group of 16 scientists (including 3 physicians) who are selected by NLM. The panel meets three times a year to examine new or revised HSDB records for accuracy and completeness. Each data statement passing this process is labeled as peer-reviewed. Data in HSDB are also labeled with an unreviewed tag. This information includes data that do not lend themselves to review, such as industry data (NLM, 1994).

Chemical Carcinogenesis Research Information System

The National Cancer Institute's (NCI) database, the Chemical Carcinogenesis Research Information System (CCRIS), provides test results from scientifically-evaluated carcinogenicity, mutagenicity, tumor production, and tumor inhibition studies.

Organization and Content

The CCRIS database is organized by chemical record and contains information on more than 7,100 chemicals. Each record has two information sections.

The substance identification section of the record includes the CAS Registry Number, as well as a classification of the chemical's major use (e.g., chelating agent, refrigerant, or surfactant) for each chemical with a commercial application. The most extensive section of the record contains the information on the scientific studies done on carcinogenicity, tumor promotion, mutagenicity, and tumor inhibition. Multiple studies are listed for each outcome. For each study, information is provided on the species, strain, and sex of the animal (data from human epidemiologic studies are included), the route and dose of exposure, the target tissue and type of lesion (as applicable), the test results, and the bibliographic reference. The CCRIS database brings together the major scientific studies on each chemical and gives extensive information on the study and the study results.

Review and Updates

Sources for CCRIS are selected by NCI scientists from the primary literature, special reviews, NCI reports (e.g., NCI/National Toxicity Program Carcinogenesis Technical Reports and NCI Short Term Test Program reports), and other authoritative sources (e.g., IARC monographs and EPA's GENE-TOX Program Reports). Studies must contain clearly positive or negative results and meet criteria for being adequately conducted. Monthly updates are performed, primarily to add additional information to existing records.

GENE-TOX

The GENE-TOX database, sponsored by EPA, contains data on mutagenicity testing for more than 2,900 chemicals. In 1979, EPA began a multiphase effort to review and evaluate the scientific literature on methodologies for genetic toxicology. In the first phase of this effort (1979–1986), 196 scientists from government, academia, and industry selected 23 assays and evaluated the published scientific literature on studies in which those assays were used (Auletta et al., 1991; Waters, 1994). Review articles were published by the work groups detailing the results of the literature review and assay evaluation. Phase II of the project updated the literature reviews for selected mutagenicity assays (e.g., the Ames assay) and established and evaluated the GENE-TOX database, which provides the results of the evaluations. Because of the costs involved, it was decided to include only the qualitative results and not the extensive supporting documentation (Waters, 1994). Phase III continues to update the information and to add data on selected new assays. As expected, assays such as the Ames/Salmonella test have been conducted with numerous chemicals (more than 2,000), whereas some bioassays have been conducted with fewer than 10 compounds (Lu and Wassom, 1992).

Organization and Content

The GENE-TOX database is organized by chemical record and is maintained by EPA. Each record contains two major sections of data: substance identification (including CAS Registry Number, chemical name, and synonyms) and results of mutagenicity studies.

For each chemical, details are given for the mutagenicity studies reviewed by the GENE-TOX panels. The information on each study includes the assay type, species and cell type, dose-response data, study results, and references to the GENE-TOX expert panel report for this assay. Additionally, the reference field specifies the five-digit number that can be searched in the EMIC database (see the description of EMIC below) to retrieve the complete citation. References for the particular study in the primary literature are also provided.

Review and Updates

As discussed above, all GENE-TOX records have received extensive evaluation. The database is updated when new peer-reviewed data become available from EPA.

Integrated Risk Information System

The Integrated Risk Information System (IRIS) provides EPA with health risk and regulatory information on more than 660 chemicals including carcinogenic and noncarcinogenic risk assessments for oral and inhalation routes of exposure. The database was initially created by EPA in 1986 as an internal resource that could be used to develop and access agency-wide EPA consensus judgments on the human health effects of chemicals (Tuxen, 1992). The database was added to the TOXNET system in 1992.

Organization and Content

Each IRIS database record has information on one chemical, which is then organized into eight major categories of data. The chemical identification fields allow the database to be searched by chemical name, CAS Registry Number, chemical synonyms, or molecular formula.

The categories of carcinogenic and noncarcinogenic assessment provide extensive technical information describing animal and human studies on each chemical. The noncarcinogenic fields provide reference doses for oral and inhalation exposures to the chemical. Reference doses are estimates of the daily amount of a chemical that humans can be exposed to throughout their lives without the risk of suffering adverse effects (NLM, 1994). The carcinogenicity assessment category describes the unit risk (the upper-level lifetime risk of contracting cancer when exposed to specific concentrations of the chemical). Carcinogenicity assessments are provided, when available, for both oral and inhalation exposures. EPA currently classifies the carcinogencity of a chemical into one of the following five classification codes: A (human carcinogen), B (probable human carcinogen), C (possible human carcinogen), D (not classifiable as to human carcinogenicity), and E (no evidence of carcinogenicity for humans) (NLM, 1994). Both the noncarcinogenic and carcinogenicity assessment categories provide descriptions of studies, references to the EPA source document for the assessment and other studies cited, the date that the data were most recently reviewed, and an EPA contact person.

Additionally, IRIS contains the advisories from the EPA Office of Drinking Water and exposure standards and regulations including the requirements of the Clean Air, Clean Water, and Safe Drinking Water Acts.

Review and Updates

IRIS is updated monthly by EPA. Updates include the addition of new records, revisions of existing records, or record deletions. All information in the IRIS database undergoes scientific review by two agency-wide work groups of EPA scientists, the Oral Reference Dose/Inhalation Reference Concentration Work Group and the Carcinogen Risk Assessment Verification Endeavor Work Group. Information is incorporated into the IRIS database after the work group has reached consensus on the health effects or dose-response assessment (Tuxen, 1992). Each record includes the revision history field, which provides the date on which that IRIS record most recently completed EPA scientific review.

Registry of Toxic Effects of Chemical Substances

The National Institute for Occupational Safety and Health (NIOSH) has developed and updated the Registry of Toxic Effects of Chemical Substances (RTECS) database since its inception in 1971. RTECS provides toxic effects and regulatory information on more than 133,000 chemicals. Mandated by the Occupational Safety and Health Act of 1970, the original edition was called the

Toxic Substances List and included information on approximately 5,000 chemicals. RTECS was added to the NLM online system in 1977 and is also available through license agreements with NIOSH in CD-ROM and online formats from a number of database vendors. Publication of print and microfiche versions of RTECS has recently ceased.

Organization and Content

RTECS is organized by chemical record, and each record contains four major categories of information: substance identification (including the CAS Registry Number), toxicity and biomedical effects, toxicology and carcinogenicity review, and exposure standards and regulations. Like CCRIS, RTECS provides results from scientific studies and does not evaluate or summarize that information. The section of the RTECS record on toxicity and biomedical effects provides information on scientific studies on mutagenicity, carcinogenicity, skin and eye irritation, general toxicity, and reproductive effects. For each study, information is provided on the type of test or study including species of test animal; the route, dose, and duration of exposure; target tissues; study results; and brief citations. RTECS uses a controlled vocabulary developed by NIOSH to facilitate searching by standardizing the descriptions of health effects, cell types, species, etc.

The toxicology and carcinogenicity review section of the RTECS record provides relevant IARC reviews, threshold value limits, and NIOSH Recommended Exposure Limits for the specified chemical.

The final RTECS section summarizes agency standards and regulations and provides a federal program status field listing the activities of government agencies and programs (e.g., the Carcinogenesis Bioassay Program) for that particular chemical.

Review and Updates

Information in RTECS is selected under the direction of NIOSH staff from key scientific journals, government reports, and other technical documents. The data come directly from the publication and do not undergo additional review by NIOSH. The RTECS database at NLM is maintained and updated by NLM with information provided by NIOSH (NLM, 1994). RTECS is updated quarterly, a process that involves adding additional records, as well as adding and deleting information from existing records.

THE TEHIP PROGRAM 43

Toxic Chemical Release Inventory

The Emergency Planning and Community Right-to-Know Act of the Superfund Amendments and Reauthorization Act of 1986 (SARA) mandated that facilities in the United States meeting its criteria must annually report information on environmental release of chemicals (due either to routine release as part of business operations or by accident) to the EPA. The Act further specified that all the emissions information be made available in a computer file format (Bronson, 1991). The end result is the Toxic Chemical Release Inventory database (TRI), which is developed and maintained by EPA and accessed through NLM's TOXNET system.

Facilities are required to report emissions if, among other criteria, they manufacture or process more than 25,000 pounds of the chemical per year. Each facility completes and submits an EPA Form R for each chemical meeting the criteria. More than 300 chemicals are on the list for mandated reporting. Separate TRI files are available for each year beginning with 1987, and the databases can be searched individually or as a group.

Organization and Content

Each TRI record corresponds to one EPA Form R submission (one chemical from one reporting facility). The TRI94 file with data on 1994 emissions contains more than 75,000 records. The major categories of data in the TRI records are facility identification (including facility name and address), substance identification (including the CAS Registry Number and manufacturing and processing uses), and environmental release of the chemical (including the amounts of air emissions, water discharges, releases to underground injection, waste treatment, and off-site waste transfer). The geographic information available in the TRI files (including zip code, city, county, or state) make these databases a valuable resource for community and other groups interested in assessing potential environmental hazards in their region.

Similarly to HSDB, the searcher of the TRI databases has three search options: direct searching (i.e., using the command language), menu searching, or using Grateful Med. Menu searching offers screens to prompt the user through the process of developing the search strategy and is useful for the wide range of users, including the general public. Grateful Med also provides similar ease of use.

TRI data can be searched and manipulated by the database through ranging and calculating operations. A search, for example, could be performed on all facilities in Virginia with total air release of ethylbenzene greater than 2,000 pounds per year. Additionally, the TRI databases can perform a number of calculating and statistical functions (e.g, average, standard deviation, mean, median, and mode).

TRIFACTS

A companion database to TRI is TRIFACTS, a factual database with information on the health effects, ecological effects, safety, and handling of the more than 300 TRI chemicals. The information in TRIFACTS is intended to provide the lay person with summary information on the TRI chemicals. When the searcher logs onto TRIFACTS, he or she is informed that TRIFACTS summaries "should be supplemented with technical literature to answer in-depth questions."

TRIFACTS summaries are adapted from the State of New Jersey Hazardous Substance Fact Sheets. The Fact Sheets were originally mandated by New Jersey's Worker and Community Right to Know Act and were developed for all chemicals on New Jersey's Right to Know Hazardous Substance List. EPA, through an agreement with the State of New Jersey, adapted the Fact Sheets for use as an online database and added EPA ecological data to the file. The TRIFACTS database was added to the TOXNET system in 1992.

Organization and Content

The TRIFACTS database has one record per chemical for most all of the over 300-plus TRI chemicals. Within each record the major categories of data are substance identification (including CAS Registry Number), chemical and physical properties, and safety and handling (including recommendations on personal protective equipment and clothing; the DoT Emergency Guidelines for firefighters, police, or emergency workers; and the National Fire Protection Association's Hazard Classification of flammability). Another major category of data includes summaries of the chemical's human toxicity and biomedical effects (Box 2.3). Information is included on emergency medical treatment procedures and acute and chronic (including cancer and reproductive) effects of the chemical on humans. Additionally, TRIFACTS provides ecological information (acute and chronic effects on aquatic and terrestrial life) and summaries of the exposure standards and regulations (including OHSA standards and NIOSH recommendations).

THE TEHIP PROGRAM

> **BOX 2.3**
> **Excerpt from the TRIFACTS Record on Toluene**
>
> NAME   Toluene
> RN   108-88-3
> ACUTE   The following acute (short term) health effects may occur immediately or shortly after exposure to toluene:
> • Exposure can irritate the nose, throat, and eyes. Higher levels can cause you to feel dizzy, lightheaded, and to pass out. Death can occur.
> • Lower levels may cause trouble concentrating, headaches, and slowed reflexes.

Directory of Information Resources Online

The DIRLINE (Directory of Information Resources Online) database is unique among the TEHIP databases in providing directory information that not only covers the fields of toxicology and environmental health but also encompasses information resources throughout all fields of health and biomedicine. Descriptions and contact information for more than 17,000 biomedical and health-related organizations, databases, software programs, and other information resources are available through DIRLINE. This database was developed by SIS staff to provide an alternate source for answering information requests and was designed to be used by health professionals, information specialists, and the general public (NLM, 1994).

Organization and Content

The content of the DIRLINE database is merged from the following sources of directory information:
- Centers for Disease Control and Prevention National AIDS Clearinghouse
- Directory of Biotechnology Information Resources
- Health Services Research Information (National Information Center for Health Services Research)
- Maternal and Child Health Information (produced by the National Center for Education in Maternal and Child Health)
- National Institutes of Health Research Resources
- NLM (list of organizations developed by the Library of Congress)
- NLM History of Medicine Division

- ODPHP Health Information Center Databases (sponsored by the Office of Disease Prevention and Health Promotion, U.S. Department of Health and Human Services)
- Poison Control Centers (data provided by the American Association of Poison Control Centers)
- Regional Alcohol and Drug Awareness Resource Network (produced by the National Clearinghouse on Alcohol and Drug Information)
- Self-Help Clearinghouses (produced in collaboration with the Surgeon General's Initiative in Self-Help and Public Health)

Each DIRLINE record provides information on one information resource (Box 2.4). DIRLINE records include contact information (including address and telephone number), a summary description of the resource, and MeSH terms. Because of the multiple sources making up the DIRLINE database, duplicate records may be included.

BOX 2.4
DIRLINE Record for the
Association of Occupational and Environmental Clinics

SI	NLM/30011
NA	Association of Occupational and Environmental Clinics
AC	AOEC
AD	1010 Vermont Ave., NW, Suite 513, Washington, DC 20005
TEL	(202) 347-4976
TEL	(202) 347-4950 (FAX)
AB	The AOEC was established in 1987 to enhance the practice of occupational and environmental medicine through information sharing, education, and research. The growing member network now includes 39 clinics, and over 230 individuals. The AOEC aids in identifying, reporting, and preventing occupational and environmental health hazards and their effects. It encourages provision of high quality clinical services for people with work or environmentally related health problems. Its members receive reports based on databases being developed by AOEC, and on other AOEC research and conferences. The AOEC answers inquiries and holds workshops and seminars on Multiple Chemical Sensitivity (MCS). It acts as a resource for patient referrals for ATSDR, NIOSH, and others.

NOTE: SI=Secondary Source ID; NA=Name; AC=Acronym; AD=Address; TEL= Telephone Number; and AB=Abstract.

Review and Updates

DIRLINE is provided free of charge as part of NLM's efforts to increase the availability of HIV/AIDS-related information to the scientific community and the general public. Access to DIRLINE does not require an NLM user code since DIRLINE is available through NLM's Locator Internet site (telnet locator.nlm.nih.gov) in addition to its availablity as part of MEDLARS (NLM, 1995). DIRLINE is updated quarterly.

TEHIP BIBLIOGRAPHIC DATABASES

Many health professionals are familiar with NLM's bibliographic databases, particularly MEDLINE. Six of the TEHIP databases are bibliographic and provide references to the vast range of toxicology and environmental health literature. Obtaining the full text of the journal article or document may require using a variety of mechanisms. For toxicology and environmental health citations listed in MEDLINE, the full text document can be ordered through NLM's online ordering system, LOANSOME DOC. However, retrieving the full text of other citations may require utilizing other scientific libraries and resources.

TOXLINE/TOXLIT

A great diversity of scientific disciplines is involved in the fields of toxicology and environmental health. As a result, the scientific literature is dispersed among numerous journals and is indexed in a variety of sources. In 1972, NLM developed TOXLINE with the goal of having a single bibliographic database that would cover the field of toxicology (Kissman and Wexler, 1985). TOXLINE was designed to incorporate the relevant records from other indexing and abstracting sources and originally was composed of records from *Index Medicus*, Biological Abstracts, Chemical Abstracts, and International Pharmaceutical Abstracts. As shown in Table 2.4, the bibliographic records in TOXLINE are drawn from 18 sources; additionally, the companion file, TOXLIT, provides access to records from certain royalty sources, currently only Chemical Abstracts (see Chapter 6 for discussion of costs).

Records from the 18 subfiles are not significantly altered or enhanced before being imported into TOXLINE. As a result, there is no single controlled vocabulary, and searchers must use a number of synonyms and similar expressions to search TOXLINE comprehensively. Additionally, the secondary sources that make up TOXLINE overlap to some extent in their coverage (e.g., MEDLINE, BIOSIS, and Developmental and Reproductive Toxicology [DART]) and may result in duplicate, although not identical, citations in TOXLINE (NLM, 1994).

TABLE 2.4 TOXLINE Subfiles

TOXLINE Subfile	Subfile Description
Aneuploidy	Collection of bibliographic citations prepared by EMIC on numerical chromosomal abnormalities
Developmental and Reproductive Toxicology (DART)[a]	DART database on teratology and many aspects of reproductive toxicology
Environmental Mutagen Information Center[a]	EMIC database on substances tested for genotoxic activity
Environmental Teratology Information Center[a,b]	ETICBACK database covering the 1950–1989 teratology literature
Epidemiology Information System[b]	Epidemiology Information System database, developed by FDA's Center for Food Safety; citations cover literature published from 1940 to 1988 on the distribution and health effects of food contaminants
Federal Research in Progress (FEDRIP)	Subset of the FEDRIP database produced by NTIS and describing current federal research and development projects
Hazardous Materials Technical Center (HMTC)[b]	*HMTC Abstract Bulletin* on hazardous wastes, published by the Department of Army's Hazardous Materials Technical Center
International Labour Office	CIS Abstracts; toxicology-related material produced by the International Labour Office's International Occupational Safety and Health Information Centre
International Pharmaceutical Abstracts (IPA)	Subset of the IPA database on development and use of drugs; produced by the American Society of Health System Pharmacists
NIOSHTIC	Subset of the NIOSH's NIOSHTIC database on occupational safety and health literature
Pesticides Abstracts[b]	EPA publication on the epidemiological effects of pesticides; the publication was terminated in 1981
Poisonous Plants Bibliography[b]	Pre-1976 citations to literature on poisonous plants

continues

TABLE 2.4 Continued

TOXLINE Subfile	Subfile Description
RISKLINE	RISKLINE database developed by the National Chemicals Inspectorate in Sweden covers literature in toxicology and ecotoxicology
Toxic Substances Control Act Test Submissions	Reports of health and safety studies on certain chemicals submitted by industry to EPA
Toxicity Bibliography (TOXBIB)	Subset of the NLM's MEDLINE file, created monthly from SDILINE, the most recent month of MEDLINE
Toxicological Aspects of Environmental Health (BIOSIS)	Subset of the BIOSIS database (Biological Abstracts and Biological Abstracts/Reports, Reviews, and Meeting Abstracts) describing the health effects of chemical substances
Toxicology Document and Data Depository	Subset of the National Institutes of Health CRISP (Computer Retrieval of Information on Scientific Projects) database containing information on ongoing research in toxicology and related topics
Toxicology Research Projects	Subset of the NTIS Government Reports and Announcements database containing government publications on toxicology and related topics

[a]Separate database also available via the TOXNET system.
[b]Closed file. Records are no longer added to TOXLINE from these subsets.

Organization and Content

Because of the size of the database files (TOXLINE and TOXLIT files contain more than 4 million records), the pre-1981 records from TOXLINE and TOXLIT (which primarily cover the literature from 1965 to 1980) have been put into separate databases (TOXLINE65 and TOXLIT65). All four files (TOXLINE, TOXLINE65, TOXLIT, and TOXLIT65) are organized as tradi-tional bibliographic databases with fields for basic bibliographic information including author, title, source, language, publication type, and abstract (when available).

MeSH indexing terms are available only for the TOXBIB, BIOSIS, and DART subfiles of TOXLINE; additionally, the keyword field provides access to the indexing terms used by the subfile producer (e.g., FEDRIP). All of the indexing terms are used, along with terms from other fields, to provide text word searching capabilities. The CAS Registry Number is found in many of the TOXLINE records.

Additionally, there are fields for specialized information to accommodate the information from the various subfiles (e.g., the Award Type field is used in the Toxicology Research Projects and FEDRIP subfiles to indicate whether the research project is intramural, contract, fellowship, or grant).

Review and Updates

TOXLINE and TOXLIT are updated monthly, with new records added from various subfiles. More than 140,000 records are added annually and the file is rebuilt each year to provide updated MeSH indexing.

Developmental and Reproductive Toxicology

Two bibliographic databases provide citations and abstracts to the scientific literature on teratology and developmental and reproductive toxicology. The ETICBACK (Environmental Teratology Information Center Backfile) database was produced by ORNL and covers the teratology literature from 1950 to 1989. A decision to expand the database to more fully cover the developmental and reproductive toxicology literature resulted in the closing of the ETIC file and the introduction of the DART database, which covers literature from 1989 to the present. Funding constraints, however, have prevented the comprehensive coverage of the scientific literature on lactation effects and on some aspects of male and female reproductive toxicology.

Organization and Content

Both DART and ETICBACK are bibliographic databases and provide searchable fields for basic bibliographic information including author, source, language, publication type, and, when available, the abstract. Substance identification fields in both databases provide searchable access to the CAS Registry Number and other identication information. MeSH terms have been added to the DART records, although some of the MeSH searching commands (e.g., explode or pre-explode) are not yet available in the TOXNET system. More than 60 percent of DART records come from MEDLINE; the others are identified by screening the literature not covered by MEDLINE (including meeting abstracts and government reports) (NLM, 1994). ETICBACK records include a number of specialized fields including assay method, experimental conditions, and maternal effects. Both databases currently receive or have received funding from National Institute of Environmental Health Sciences (NIEHS), EPA, NLM, and the Food and Drug Administration's National Center for Toxicological Research.

Review and Updates

ETICBACK is now a closed file, meaning that no new records are being added to the file. DART is updated monthly and is developed and maintained by NLM, with additions and deletions to the database made as necessary. Each month a search profile is run against the SDILINE database (a subset of the MEDLINE database containing only the most recent month's input into MEDLINE), and relevant records are added to DART. Additionally, relevant sources not indexed by MEDLINE are added.

Environmental Mutagen Information Center

EMIC and its backfile, EMICBACK, cover the scientific literature on genetic toxicity testing. In the late 1960s increased public and scientific concern about the mutagenic actions of chemicals led to the establishment of the Environmental Mutagen Information Center (EMIC) at ORNL (Wassom and Lu, 1992). One of the major functions of the Center since its inception has been the collection and organization of the scientific literature on chemical, biologic, and physical agents that have been tested for genetic toxicity. The result is the EMIC database, which was originally produced by ORNL and which is now managed by NLM. Support for EMIC is provided by EPA and NIEHS.

Organization and Content

Similarly to TOXLINE, EMIC has been separated into two bibliographic database files. The EMICBACK file covers the pre-1950 through 1991 literature, and the EMIC file includes the literature published since 1991. The two files contain more than 88,000 citations to the genetic toxicology literature, which can also be accessed as the EMIC subfile of the TOXLINE database.

In addition to the bibliographic fields (e.g., author, title, source, and language), EMIC and EMICBACK contain a number of specialized fields regarding the assay and the study. These fields include keywords describing the test organism, tissue cultured, type of assay, and experimental conditions. EMIC records also contain fields identifying the substances involved in the test including the names of the test, the inducer, and the control agents. As described above, EMIC contains the records of sources used in EPA's GENE-TOX program, including the panel reports of the GENE-TOX reviewers.

Review and Updates

EMIC records are selected and indexed by ORNL. The EMICBACK file is closed (no new records will be added), and the EMIC file is updated monthly. Approximately 3,000 records are added to EMIC each year.

CONCLUSIONS

Although this chapter presents only an overview of each of the TEHIP databases, it is evident that there is a wealth of information available for use by health professionals, researchers, industry, policymakers, and the general public. What is also evident is the complexity of the TEHIP program—a complexity that results from the number of databases, the disparate scope and content of the databases, the diverse sources of information, and the variations in the type of information provided. The following chapters examine these issues and provide the committee's recommendations for facilitating health professionals' use of toxicology and environmental health information resources.

REFERENCES

Auletta AE, Brown M, Wassom JS, Cimino MC. 1991. Current status of the Gene-Tox Program. *Environmental Health Perspectives* 96:33–36.

Bronson RJ. 1991. Toxic Chemical Release Inventory information. *Medical Reference Services Quarterly* 10(1):17–34.

Brooks SM, Gochfeld M, Jackson RJ, Herzstein J, Schenker MB, eds. 1995. *Environmental Medicine*. St. Louis: Mosby.

Campbell-Kelly M, Aspray W. 1996. *Computer: A History of the Information Machine*. New York: Basic Books.

Kissman HM, Wexler P. 1985. Toxicology information systems: A historical perspective. *Journal of Chemical Information and Computer Sciences* 25:212–217.

Lowe HJ, Barnett GO. 1994. Understanding and using the Medical Subject Headings (MeSH) vocabulary to perform literature searches. *Journal of the American Medical Association* 271(14):1103–1108.

Lu P-Y, Wassom JS. 1992. Risk assessment and toxicology databases for health effects assessment. In: U.S. Environmental Protection Agency, Oak Ridge National Laboratory (ORNL). *Proceedings of the Symposium on the Access and Use of Information Resources in Assessing Health Risks from Chemical Exposure*. Oak Ridge, TN: ORNL.

Miles WD. 1982. *A History of the National Library of Medicine*. NIH Publication No. 85-1904. Bethesda, MD: National Institutes of Health.

Netscape. 1996. Internet time line. *Inside Netscape Navigator* 1(4):8–9.

NLM (National Library of Medicine). 1986. *Building and Organizing the Library's Collection. Long Range Plan, Report of Panel 1*. Bethesda, MD: NLM.

NLM. 1993. *Improving Toxicology and Environmental Health Information Services.* Report of the Board of Regents Long Range Planning Panel on Toxicology and Environmental Health. NIH Publication No. 94-3486. Bethesda, MD: NLM

NLM. 1994. *Reference Materials for the Toxicology Information Program Online Services.* Bethesda, MD: NLM.

NLM. 1995. *National Library of Medicine Programs and Services, 1994.* NIH Publication No. 95-256. Bethesda, MD: NLM.

NLM. 1996a. *The National Library of Medicine* [http://www.nlm.nih.gov/publications/factsheets/nlm.html]. November

NLM. 1996b. *NLM Policy on Database Pricing* [http://www.nlm.nih.gov/publications/factsheets/datapric.html]. November.

NLM. 1996c. *NLM Online Services Program Policy Statement* [http://www.nlm.nih.gov/publications/factsheets/online_serv_policy.html]. November.

NLM. 1996d. *Toxicology and Environmental Health Information Program* [http://sis.nlm.nih.gov/tehipfs.htm]. November.

PSAC (President's Science Advisory Committee). 1966. *Handling of Toxicological Information.* Washington, DC: White House.

Rom WN, ed. 1992. *Environmental and Occupational Medicine*, 2nd ed. Boston: Little, Brown, and Company.

Siegel ER, Cummings MM, Woodsmall RM. 1990. Bibliographic retrieval systems. In: Shortliffe EH, Perreault LE, eds. *Medical Informatics: Computer Applications in Health Care.* Reading, MA: Addison Wesley.

Tesler LG. 1991. Networked computing in the 1990s. *Scientific American* 265(3):86–93.

Tuxen L. 1992. Integrated Risk Information System. In: U.S. Environmental Protection Agency, Oak Ridge National Laboratory (ORNL). *Proceedings of the Symposium on the Access and Use of Information Resources in Assessing Health Risks from Chemical Exposure.* Oak Ridge, TN: ORNL.

Van Camp AJ. 1989. The TOXNET gateway. *Online* July:70–74.

Wassom JS, Lu P-Y. 1992. Evolution of toxicology information systems. In: U.S. Environmental Protection Agency, Oak Ridge National Laboratory (ORNL). *Proceedings of the Symposium on the Access and Use of Information Resources in Assessing Health Risks from Chemical Exposure.* Oak Ridge, TN: ORNL.

Waters MD. 1994. Development and impact of the Gene-Tox Program, genetic activity profiles, and their computerized data bases. *Environmental and Molecular Mutagenesis* 23(Suppl. 24):67–72.

Zenz C, Dickerson OB, Horvath EP, eds. 1994. *Occupational Medicine.* St. Louis: Mosby.

3

Other Toxicology and Environmental Health Information Services

The TEHIP databases represent only a small subset of the numerous databases containing information related to toxicology and environmental health. This chapter discusses some of the other toxicology and environmental health information resources currently available and describes a role for NLM in providing access to this information.

Responsibilities for research, regulation, and risk communication on environmental health issues are fragmented between numerous local, state, and federal government agencies, international organizations, industry, and other private-sector businesses. As a result, numerous databases that are relevant to fulfilling each organization's specific environmental health mission or goals have been developed. As seen in Figure 3.1, responsibility for the federal government's involvement in environmental health spans numerous jurisdictions and boundaries, including most federal departments and many agencies. The cross-cutting nature of environmental health issues can be seen in the range of concerns that federal agencies are mandated to address, for example:

- **Populations and individuals exposed to potential hazards through occupational, environmental, and accidental exposures.** Responsible agencies include the U.S. Departments of Defense (DoD), Energy (DoE), Health and Human Services (DHHS), Labor, Veterans Affairs, the Environmental Protection Agency (EPA), and the Federal Emergency Management Agency.

- **Manufacture, use, transportation, treatment, storage, and disposal of hazardous chemicals.** Responsible agencies include the U.S. Departments of

Commerce (DoC), DoD, DoE, Transportation (DoT), EPA, and the Consumer Product Safety Commission.

- **Exposure pathways** (including air, water, and soil). Responsible agencies include the U.S. Departments of Agriculture (DoA), Interior (DoI), DoC, and EPA.

Additionally, environmental health concerns are often specific to a localized region or a particular population, because, for example, of a chemical spill (e.g., Superfund sites) or an occupational exposure. As a result significant sources of data at the state and local levels are incorporated into databases. Many authoritative international sources of toxicology and environmental health information are also available, including the International Agency for Research on Cancer, the International Labour Office's Occupational Safety and Health Information Centre, and the United Nations Environmental Programme. The private sector is also involved in the development of online factual and bibliographic databases related to toxicology and environmental health.

Each agency, organization, or business collects and organizes information specific to its mission and develops online databases each with its own focus, search language, and unique database fields and input methodologies. Generally, there is no standardization of data collection, storage, analysis, retrieval, or reporting of environmental health data across the federal government or between the private and public sectors (Sexton et al., 1992). Database software and search interfaces use diverse computer operating systems that are frequently incompatible (Lu and Wassom, 1992). Thus, the challenges to health professionals and other interested users of environmental health information are, first, to be aware of and locate the database(s) that contains the information to address their question; second, to have the proper computer connection to the database(s); and finally, to understand the background and nature of the information, including the implications of data collection methodologies.

Table 3.1 lists only a sample of the toxicology and environmental health databases available. It focuses primarily on federal government databases and is not a comprehensive list, but rather serves to illustrate the number and diversity of information resources available in this field (see also EPA et al., 1992; Wexler, 1988). Additionally, it should be noted that there are a rapidly expanding number of Internet Web sites that compile information on this field and that aim to provide information to diverse audiences including advocacy groups and the general public.

It is not within the scope of this report to address the plethora of issues that would be involved in attempting to coordinate the development and management of environmental health databases at the federal level or beyond. The committee is aware of several ongoing coordination efforts within the federal government. The DHHS Data Council is addressing issues relevant to the coor-

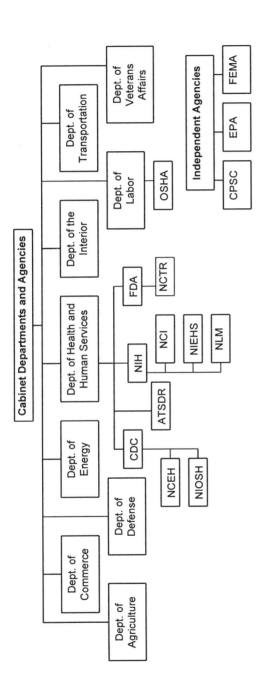

FIGURE 3.1 Executive branch departments and agencies involved in environmental health issues. NOTE: ATSDR=Agency for Toxic Substances and Disease Registry; CDC=Centers for Disease Control and Prevention; CPSC=Consumer Product Safety Commission; EPA=Environmental Protection Agency; FDA=Food and Drug Administration; FEMA=Federal Emergency Management Agency; NCEH=National Center for Environmental Health; NCI=National Cancer Institute; NCTR=National Center for Toxicological Research; NIEHS=National Institute of Environmental Health Sciences; NIH=National Institutes of Health; NIOSH=National Institute for Occupational Safety and Health; NLM=National Library of Medicine; OSHA=Occupational Safety and Health Administration.

dination of health and nonhealth data collection and data analysis activities within DHHS, including coordinating health data standards.

Additionally, DHHS has an interagency Environmental Health Policy Committee (EHPC) that focuses on the coordination of environmental health policy and programs. There are liaison members to EHPC from six other federal agencies and departments. One subcommittee of EHPC is specifically examining environmental health information issues. Currently, NLM has a representative on EHPC, but NLM is not represented on the subcommittee on environmental health information issues. The committee believes that is important for NLM, and specifically, the staff of the TEHIP program, to be involved in these coordination efforts, particularly those focused on information issues.

One model for coordination efforts is the ongoing development and implementation of the National Environmental Data Index (NEDI). NEDI is an element of the National Information Infrastructure and is designed to provide distributed access to existing environmental information systems with the goal of facilitating access to this information by the general public, the scientific community, government, and industry (NEDI, 1996). The National Oceanic and Atmospheric Administration (NOAA) is leading the implementation of NEDI on the Internet with the cooperative efforts of DoA, DoC, DoD, DoE, and DoI, as well as EPA, the National Aeronautics and Space Administration (NASA), and the National Science Foundation (see Chapter 7).

CONCLUSION AND RECOMMENDATION

Although the committee's charge did not include the compilation of information on all of the toxicology and environmental health information resources, in assembling Table 3.1, the committee noted the vast extent of available information in toxicology and environmental health, as well as its utility to health professionals. Unfortunately, because of the diffuse sources of environmental health information and the disparate manner in which data are collected, the multitude of databases may not be known or easily available to those who would benefit from this information.

Thus, the committee believes that it is important for NLM to carry out the traditional and expert role of a library by organizing (cataloging) online information resources in toxicology and environmental health beyond the TEHIP databases and increasing health professionals' and other interested users' awareness of the relevant resources. Comparable to cataloging books and journals on toxicology and environmental health, the committee believes NLM's role should include a cataloging of databases and other online information resources in this field. By providing users with information on non-NLM resources (e.g., a description of the information resource and its access points),

NLM will be delivering the valuable library service of providing users with the information needed to access the most relevant resource available. One ongoing project with this emphasis is TEHIP's Internet World Wide Web page with links to other toxicology and environmental health-related information resources both within the federal government and internationally. This emphasis on the broad spectrum of information resources in toxicology and environmental health could also be included as an integral part of the TEHIP program's training and outreach activities (Chapter 5), making these activities an education on the realm of information resources in this area. In order to carry out this role effectively, it will be necessary to incorporate an evaluation mechanism and to consider the funding required to implement this recommendation.

The committee recommends that NLM consider expanding its traditional library services in toxicology and environmental health by organizing and cataloging the full spectrum of online toxicology and environmental health information resources.

TABLE 3.1 Sample of Current Toxicology and Environmental Health Databases

Source	Database	Subject Content
	Federal Government Databases	
Department of Agriculture		
National Agricultural Library	Agricola	Bibliographic database covering scientific literature on general agriculture, animal and plant science, aquatics, and pollution
EPA	ACQUIRE (Aquatic Toxicity Information Retrieval System)	Factual and full-text database of toxic effects of approximately 5,100 chemicals on 2,400 aquatic organisms and plants
	CERCLIS (Comprehensive Environmental Response, Compensation, and Liability Information System)	Directory of information on >36,000 releases of hazardous substances and Superfund data on hazardous waste site assessment and remediation
	Dermal Absorption	Bibliographic, factual, and full-text database of qualitative and quantitative health effects of approximately 3,000 chemicals administered via the dermal route

continues

TABLE 3.1 Continued

Source	Database	Subject Content
	ENVIROFATE (Environmental Fate)	Factual and full-text database on the environmental fate of approximately 800 chemicals (produced in quantities exceeding 1 million pounds per year) released into the environment
	Gastrointestinal Absorption Database (GIABS)	Bibliographic database covering the literature on experiments in gastrointestinal absorption, metabolism, and excretion
	IRIS (Integrated Risk Information System)[a]	Factual database on health and regulatory information on >660 chemicals; includes carcinogenic and noncarcinogenic risk assessment data
	Oil and Hazardous Materials/Technical Assistance	Information on the assessment of hazards encountered due to oil discharges or hazardous substance spills
	Resource Conservation and Recovery Information System	Directory, factual, and full-text database of information on hazardous waste generation and management facilities and transporters
	TRI (Toxic Chemical Release Inventory)[a]	Factual database with data on the release and transfer of >300 chemicals by medium and site of release (air, water, underground, land, off-site)
EPA	TRIFACTS[a]	Factual database containing information related to health and ecological effects, safety, and handling information on most chemicals listed in the TRI database
	TSCA (Toxic Substances Control Act) Chemical Substances Inventory	Directory, factual, and full-text database of information on approximately 131,000 chemicals used in commerce in the United States and covered by the TSCA Initial Inventory
	TSCA Test Submissions	Bibliographic, factual, and full-text database on approximately 4,200 chemical substances described in health and safety reports submitted by chemical manufacturers, users, and importers
EPA/IARC	EPA/IARC Genetic Activity Profiles	Factual database includes genetic activity profiles on >300 agents
EPA/DoT	ERNS (Emergency Response Notification System)	Factual database with information on >275,000 initial notifications of oil discharges and hazardous substances releases in the United States since 1986

continues

TABLE 3.1 Continued

Source	Database	Subject Content
FEMA	GEMS (Global Emergency Management System)	An international index of emergency management resources compiled and maintained by FEMA; includes access to information on emergency management, fire fighting, search and rescue, etc., through Internet links to agencies with the required expertise
Department of Commerce		
NOAA	NEDI (National Environmental Data Index)	Index of environmental data and information descriptions related to human health, safety, and welfare across several federal agencies; information from NOAA, DoE's Office of Scientific and Technical Information, United States Geological Survey, NASA, EPA, and others
Department of Energy		
Oak Ridge National Laboratory	EMIC (Environmental Mutagen Information)[a]	Bibliographic database on the testing of chemicals and other agents for mutagenicity and genetic toxicology
Office of Epidemiologic Studies	PAREP (Populations at Risk to Environmental Pollution)	Studies of off-site populations who are at risk for environmental pollution
	ICIE (Information Center for Internal Exposure)	Information on doses of radionuclides and body burden
	CEDR (Comprehensive Epidemiologic Data Resource Program)	Epidemiologic information from the Health and Mortality Study of the DoE workforce
	Human Radiation Experimentation Data Base	Information from DoE files on radiation exposures
Department of Health and Human Services		
CDC	CDC Wonder	General-purpose public health information and communication system that provides query access to about 40 text-based and factual databases
ATSDR	HazDat	Toxicological profiles for public health assessments and studies on various sites related to environmental and occupational exposure to hazardous chemicals

continues

TABLE 3.1 Continued

Source	Database	Subject Content
NCI	CANCERLIT	Bibliographic database covering all aspects of cancer
NCI	CCRIS (Chemical Carcinogenesis Research Information System)[a]	Bibliographic, factual, and full-text database containing the results of carcinogenicity, tumor promotion, and mutagenicity studies on >7,100 chemicals
	PDQ	Directory and full-text database of information summaries on treatment (for health professionals and for patients), supportive care, screening and prevention, and investigational or newly approved drugs for cancer
NCI/NIEHS	National Toxicology Project (NTP) Genetic Toxicity Database	Results of long-term carcinogenicity studies with mice and rats
NIEHS/NTP	Chemical Health and Safety Data	Information on >2,000 chemicals studied by NTP related to health and safety
NIOSH	NIOSHTIC (NIOSH Technical Information Center)	Bibliographic database covering occupational safety and health literature
	RTECS (Registry of Toxic Effects of Chemical Substances)[a]	Factual database on >133,000 chemicals; contains toxicity data including threshold limits, carcinogenesis bioassay results, and regulation status
NLM/EPA/NIEHS	DART (Developmental and Reproductive Toxicology)[a]	Bibliographic database covering literature on teratology and other aspects of developmental and reproductive toxicology; contains citations to literature published since 1989 (teratology literature from 1950 to 1988 can be found in ETICBACK)
NLM	ChemID[a]	Factual and full-text database containing a chemical dictionary information on >294,000 compounds; includes Superlist information on regulatory lists
	CHEMLINE (Chemical Dictionary Online)[a]	Chemical dictionary file for >1.4 million chemicals; includes CAS Registry Numbers, chemical synonyms, molecular formulas, chemical names, and other identifiers

continues

TABLE 3.1 Continued

Source	Database	Subject Content
	HSDB (Hazardous Substances Data Bank)[a]	Factual database on toxicology of >4,500 hazardous chemicals; includes information on emergency handling, treatment, detection, environmental fate, and regulatory requirements
	TOXLINE and TOXLIT[a]	Bibliographic databases with toxicology-related citations from MEDLINE and 17 other sources including Chemical Abstracts, BIOSIS, and NIOSHTIC
Department of Defense		
Office of the Under Secretary of Defense for Acquisition and Technology	DENIX (Defense Environmental Network and Information Exchange)	Bulletin board with information on environmental legislation, compliance, restoration, clean-up, safety and occupational health, security, and DoD guidance information; provided to DoD personal and contractors in the environmental security arena
Department of Transportation		
Coast Guard	CHRIS (Chemical Hazards Response Information System)	Information on >1,210 chemicals for use in spill situations; full-text and factual database

International Databases

Source	Database	Subject Content
Canada		
Canadian Centre for Occupational Health and Safety	CHEMINFO/ INFOCHIM	Factual and full-text database on health, occupational control measures, storage, and handling information on chemicals and chemical mixtures
	MSDS/FTSS	Factual and full-text database on material safety data sheets for commercially-available chemical substances
Commission of the European Communities	ECDIN (Environmental Chemicals Data and Information Network)	Factual database on >120,000 chemicals; includes information on acute and chronic toxicity, first-aid treatment, carcinogenicity, mutagenicity, relevant legislation, and civil protection
IARC	IARCancerDisc	Factual and full-text database containing complete text of IARC reports on exposures to environmental chemicals and human cancer
Sweden		
National Swedish Board of Occupational Safety and Health	Amilit	Research reports on occupational safety and health

continues

TABLE 3.1 Continued

Source	Database	Subject Content
National Chemicals Inspectorate Library, Sweden	RISKLINE	Bibliographic database on risk assessment reports in toxicology and ecotoxicology
United Kingdom		
Royal Society of Chemistry	CSNB (Chemical Safety Newsbase)	Bibliographic database on occupational hazards involving chemicals
Great Britain Health and Safety Executive	HSELINE	Bibliographic database of the worldwide literature on occupational safety and health
	DHSS-NSF Health Aspects of Food and Environment	Bibliographic citations to literature on safety of food, water, and the environment
United Nations		
International Labor Office	CIS DOC (CIS Abstracts or CISILO)	Bibliographic citations to worldwide literature on occupational safety and health
UNEP	International Register of Potentially Toxic Chemicals	Factual and full-text database with chemical, toxicological, environmental, and legal data on hazardous chemicals
UNEP and the International Programme on Chemical Safety	Chemicals Currently Being Tested for Toxic Effects	Chemicals currently being tested for toxic effects (other than carcinogenicity)
	GELNET (Global Environmental Library Network)	Links libraries serving the members of the GEENET network in which each library has established an Environmental Health Reference Center containing essential information on environmental health hazards and their control
Other Databases		
AEA Technology	MHIDAS	Factual and full-text database containing information on >6,000 incidents involving hazardous materials that resulted in or had the potential to produce an impact "off-site."
BIOSIS	TOXBIO	Bibliographic citations to literature on toxicology and related issues; derived from BIOSIS Previews
	TOXCAS	Bibliographic citations to chemical literature on toxicology and related issues; derived from Chemical Abstracts

continues

TABLE 3.1 Continued

Source	Database	Subject Content
Cambridge Scientific Abstracts	Health and Safety Science Abstracts	Bibliographic citations on safety science and hazard control with topics on industrial and occupational health
	Pollution Abstracts	Bibliographic citations on environmental pollution research and the toxicology of pesticides, radiation, and other occupational and environmental hazards from pollution
	Toxicology Abstracts	Bibliographic citations on toxicology and on clinical and environmental toxicology
Chemical Abstracts Service	Chemical Abstracts	Bibliographic database covering the scientific literature on chemicals
Chemical Information Systems, Inc.	Clinical Toxicology of Commercial Products (CTCP)	Factual database with chemical and toxicological information on > 20,000 commercial products
Environmental Data Resources (EDR), Inc.	EDR Combined State Environmental Records	Directory of information on state hazardous waste sites, solid waste facilities and landfills, leaking underground storage tanks, and registered underground storage tanks
	EDR Superfund Material	Full-text and directory information on Superfund clean-up priority sites, including related technical and health issues
Elsevier Science	EMTOX	Bibliographic citations to literature on drug toxicity and environmental toxicology
Michigan Department of Natural Resources	CESARS (Chemical Evaluation Search and Retrieval System)	Factual databases on chemical toxicity of approximately 370 chemicals to humans, animals, and aquatic and plant life; physical-chemical properties; and environmental fate
Micromedex, Inc.	POISINDEX	Full-text database on substance identification and on management and treatment protocols for >650,000 toxic and nontoxic substances
Micromedex, Inc.	TOMES PLUS	Bibliographic, factual, and full-text database containing chemical, medical, and toxicology information; includes clinical effects, range of toxicity, and workplace standards
Occupational Health Services	EHN (Environmental Health News)	Full-text of news stories and regulations related to environmental and occupational health

continues

TABLE 3.1 Continued

Source	Database	Subject Content
	HAZARDLINE	Factual database with regulatory, health, and precautionary data on >90,000 hazardous chemicals
	OHS Material Safety Data Sheets	Full-text, directory, and factual database on >90,000 chemical substances including their toxicity, health effects, first aid, and antidotes
Purdue University	National Pesticide Information Retrieval System	Factual and full-text database of information on the active ingredients in approximately 60,000 pesticide products
Reproductive Toxicology Center, Washington, DC	REPROTOX	Bibliographic and full-text database on effects of >4,000 drugs and industrial and environmental chemicals on human reproduction
Resources Consultants, Inc.	CHEMTOX Online	Factual database of toxicology profiles and emergency management information on >10,300 regulated chemicals
Technical Database Services (TDS), Inc.	Carcinogenicity Information Database of Environmental Substances (CIDES)	Bibliographic and factual database on test results of carcinogenic and mutagenic effects of approximately 1,000 substances
University of Washington	TERIS	Factual database containing summaries of risk of teratogenic effects

NOTE: ATSDR=Agency for Toxic Substances and Disease Registry; CDC=Centers for Disease Control and Prevention; DoD=U.S. Department of Defense; DoE=U.S. Department of Energy; EPA=U.S. Environmental Protection Agency; FEMA=Federal Emergency Management Agency; IARC=International Agency for Research on Cancer; NASA=National Aeronautics and Space Administration; NCI=National Cancer Institute; NIEHS=National Institute of Environmental Health Sciences; NIOSH=National Institute for Occupational Safety and Health; NLM=National Library of Medicine; NOAA=National Oceanic and Atmospheric Administration; and UNEP=United Nations Environmental Programme.

[a]Database is a part of the TEHIP program.

SOURCES: ASTDR (1996), CDC (1996), DoD (1996), EPA (1995), FEMA (1996), NCI (1996), NEDI (1996), NIEHS (1996), NIOSH (1996), NLM (1996), Nolan, (1995), Springer-Verlag Heidelberg (1996), University of Washington (1996), WHO (1996).

REFERENCES

ATSDR (Agency for Toxic Substances and Disease Registry). 1996. *ATSDR's Hazardous Substances Release/Health Effects Database* http://atsdr1.atsdr.cdc.gov:8080/hazdat.html#A3.1]. November.

CDC (Centers for Disease Control and Prevention). 1996. *CDC Wonder* [http://wonder.cdc.gov]. November.

DoD (U.S. Department of Defense). 1996. *Denix: Defense Environmental Network and Information Exchange* [http://denix.cecer.army.mil/denix/denix.html]. November.

EPA (Environmental Protection Agency). 1995. *An Overview of ERNS*. Washington, DC: Office of Solid Waste and Emergency Response, EPA.

EPA, Centers for Disease Control, Agency for Toxic Substances and Disease Registry. 1992. *Inventory of Exposure-Related Data Systems Sponsored by Federal Agencies*. EPA Publication No. EPA/600/R-92/078. Prepared by Eastern Research Group, Inc., Lexington, MA.

FEMA (Federal Emergency Management Agency). 1996. *Global Emergency Management System* [http://www.fema.gov]. November.

Lu P-Y, Wassom JS. 1992. Risk assessment and toxicology databases for health effects assessment. In: Oak Ridge National Laboratory (ORNL), U.S. Environmental Protection Agency. *Proceedings of the Symposium on the Access and Use of Information Resources in Assessing Health Risks from Chemical Exposure*. Oak Ridge, TN: ORNL.

NCI (National Cancer Institute). 1996. *CancerNet* [http://www.nci.nih.gov/clinical/cnet.htm]. November.

NEDI (National Environmental Data Index). 1996. *National Environmental Data Index* [http://esdim.noaa.gov]. November.

NIEHS (National Institute of Environmental Health Sciences). 1996. *National Toxicology Program Chemical Health and Safety Data* [http://ntp-db.niehs.nih.gov/Main_pages/Chem-HS.HTML]. November.

NIOSH (National Institute for Occupational Safety and Health). 1996. *RTECS* [http://www.cdc.gov/niosh/rtecs.html]. November.

NLM (National Library of Medicine). 1996. *NLM Online Databases and Databanks* [http://www.nlm.nih.gov/publications/factsheets/online_databases.html]. November.

Nolan KL, ed. 1995. *Gale Directory of Databases*, Vols. 1 and 2. New York: Gale Research Inc.

Sexton K, Selevan SG, Wagener DK, Lybarger JA. 1992. Estimating human exposures to environmental pollutants: Availability and utility of existing databases. *Archives of Environmental Health* 47(6):398–407.

Springer-Verlag Heidelberg. 1996. *ECDIN CD-ROM* [http://www.springer.de/newmedia/chemist/gg/ecdin.htm]. November.

University of Washington. 1996. *TERIS: Teratogen Information System* [http://weber.u.washington.edu/~terisweb/teris/index.html]. November.

Wexler P. 1988. *Information Resources in Toxicology*, 2nd edition. New York: Elsevier.

WHO (World Health Organization). 1996. *GEENET Home Page* [http://who.unep.ch/geenet]. November.

4

Understanding the Information Needs of Health Professionals

As knowledge about the health effects of exposure to occupational and environmental chemicals increases, health professionals and other interested individuals need to be able to access and use resources that provide timely and accurate toxicology and environmental health information in an efficient and accessible manner. To better understand the toxicology and environmental health information needs of health professionals and the methods by which they locate this information, the committee reviewed the published literature (which deals primarily with physicians and their general methods of seeking information) and focused on those issues specific to toxicology and environmental health information by receiving input from health professionals through focus group discussions (see Appendix C), data collected from the questionnaire (see Appendix B), and committee discussions with colleagues and other health professionals.

INFORMATION NEEDS

In 1967, R.E. Maizell estimated that the half-life of the current information known by scientists, engineers, and health professionals was close to 10 years; that is, in 10 years half of what is learned by those in technical fields will become obsolete, and half of what such professionals will need to know is not yet available (Maizell, 1967). Advances in technology continue to increase the expansion of biomedical information at such a rapid rate that health professionals cannot possibly absorb and retain all of the information available. However, the development of online information resources, such as the TEHIP databases, have the potential to alleviate this problem by providing a framework for storing, processing, and retrieving needed information. It then becomes important to fully understand the information needs of health professionals so that the correct information is collected and stored in an accessible manner. Thus, the ability to define the information needs of health professionals (although not well-studied) is essential to the development of systems that will support their needs.

Why Health Professionals Need Information

The information needs of health professionals stem from a variety of factors including patient care, patient education, professional curiosity, and research. Additionally, the rapid advancements in technology and science have expanded the knowledge base in all fields of medicine and health care.

In 1991, Osheroff and colleagues developed a typology that assesses the information needs of health professionals, specifically physicians, by analyzing the questions posed during clinical teaching. They concluded that information needs are driven by the extent of a patient's problem, a patient's inquiry, the professional's knowledge base, and his or her level of awareness of available resources and curiosity to find out more information. The study defined the information needs of health professionals in terms of three components: (1) information that is needed for decision making and that is already known by the health professional (currently satisfied needs); (2) information that is not known by the health professional but that he or she recognizes as being applicable to the decision-making process (consciously recognized needs); and (3) information that is important to the circumstances at hand but that the health professional does not realize is applicable (unrecognized needs) (Osheroff et al., 1991).

Williams and colleagues (1992) categorized the information needs of health professionals by the reason that health professionals begin their search for information, to:

- confirm or disconfirm existing knowledge;
- assist in solving a new or unfamiliar health care problem;
- update basic knowledge on a topic through review;
- obtain information from another specialty when dealing with a patient or person with multiple problems;
- highlight particular patient care concerns to other members of the health care team;
- find out about a rare or unusual patient care problem;
- determine whether a knowledge gap exists in the literature and whether a new research project or publication should be planned; or
- assist in implementing new administrative or organizational initiatives.

Information Needed by Health Professionals

The type of information needed by health professionals is dependent on many factors, including the topic or issue in question, the knowledge base of the health professional, his or her awareness of information sources, the associated costs of acquiring information, and the purpose for which the information will be used. The "trigger" or starting point for conducting a search for information varies widely among health professionals. Focus group participants indicated that most of their inquiries related to environmental health information begin with the name of a chemical substance or with symptoms potentially associated with a known exposure (Appendix C). For example, emergency room and primary care professionals stated that they often need to "translate" the brand name of a household product into the scientific chemical name before beginning a common starting points included the place of work, type of job, or geographic location (see also Chapter 6).

A 1988 IOM committee examined the growing occupational and environmental information needs of health professionals, particularly of primary care physicians, and formulated a list of information search. Other needs that included: causative agents of occupational and environmental illnesses; signs, symptoms, and diagnosis of and treatment for occupational and environmental illnesses; nonclinical and supportive interventions; and disease and exposure patterns within the community (IOM, 1988).

Once an information need has been identified, determining who actually conducts the information search depends on a number of factors, including having the ability to conduct a database search (access to the database and the knowledge required to conduct the search); having time available to conduct the search; and determining whether a high value is placed on seeing the data and information available and following leads and data "trails." Focus groups indicated that specialists and researchers in toxicology and environmental health, as well as information specialists, public health officials, educators, and students, are more likely to conduct searches themselves, whereas physicians in general or health professionals working in nontoxicology-related fields are most likely to have others conduct a search for them.

FACTORS AFFECTING INFORMATION SEEKING

When health professionals make a decision on whether to pursue an information query, they must weigh and reach a compromise between a number of conflicting factors including the need for the information and the associated costs of locating the information such as the time, effort, and financial costs involved in information seeking (Connelly et al., 1990). Huth (1985) analyzed information-seeking methods in terms of a utility-cost analysis. In this approach (as seen below), the value of the information resource depends on the utility of the retrieved information (its relevance, thoroughness, and the efficiency with which it is retrieved) as it relates to the costs necessary to obtain the information (including the purchase cost and time and other access costs).

$$\text{Value} = \frac{\text{relevance} + \text{thoroughness} + \text{efficiency}}{\text{cost}}$$

Studies on the information-seeking behaviors of health professionals have found that colleagues and reference books are frequently the information resources that health professionals first turn to with an information query (Blackwelder and Dimitroff, 1996; Covell et al., 1985; Curley et al., 1990; Williamson et al., 1989). The access costs are minimal, and the relevance and efficiency are usually high. A recent study of nurses, physicians, pharmacists, occupational and physical therapists, and other health professionals found that 45 percent would first consult a colleague to answer an information query, 28 percent would first use personal files and reference collections, and 17 percent would first consult a librarian (Blackwelder and Dimitroff, 1996). Similarly, a 1990 survey revealed that nurses sought information from colleagues 45 percent of the time, consulted written sources 45 percent of the time, and used other resources including databases 10 percent of the time (Corcoran-Perry and Graves, 1990). Participants in the focus group discussions (see Appendix C) identified similar factors influencing the way in which they locate information. A number of studies have shown that owing to a host of factors, including lack of awareness, access, training, and time, many health professionals have not developed effective and efficient information-seeking habits that expand beyond colleagues and textbooks (Covell et al., 1985; Haynes et al., 1990; Williamson et al., 1989; Woolf and Benson, 1989). Studies of younger health professionals who have consistently used computers during their education and training show that they are more likely to search the medical literature online, whereas many older professionals are more likely to be unfamiliar with searching online databases and tend to turn to colleagues and textbooks for information (Gruppen, 1990; King, 1987; Lockyer et al., 1985; Osiobe, 1985).

The following sections discuss a number of the factors that influence information-seeking behaviors in health professionals. An understanding of these factors is critical for the development of future information resources that best serve the diverse information needs of health professionals.

Demographics

Although current health professionals appear to be in a transition phase regarding computer use, the new generation of young health professionals is more computer literate and more reliant on computers for many different kinds of information. Many current medical and nursing students spent their high school and college careers using computers for school and recreation. School libraries at all levels have online library catalogs, and the use of computers in schools has been increasing rapidly.[1] In the next 10 years, there will be a dramatic shift in student computer expertise, in computer familiarity, and in the assumption on the part of students that information is more easily located online than in reference books. This changing dynamic will have a significant impact on computer utilization skills and on the extent of online database searching.

Work or Practice Setting

The work environment or practice setting also affects the information-seeking behaviors of health professionals. There are wide variations between work settings in the opportunities that health professionals have to access online databases and computer networking capabilities. Health professionals in smaller work environments or practice settings (e.g., clinics and private practices) tend to rely more heavily on informal sources, such as colleagues, and limited formal sources, including textbooks and handbooks (Dalrymple, 1990). Although health professionals working in academic settings or in larger clinics and hospitals associated with medical schools or research institutions still refer to textbooks and colleagues frequently, they are more likely to have access to health science libraries and extensive computer networks (Dalrymple, 1990; Gruppen, 1990; Osiobe, 1985). As computers continue to become more commonplace in all health care settings, access to online resources will no longer be an issue of concern.

Time Sensitivity and Level of Detail

The time sensitivity factor, or how quickly an answer is required, is closely correlated to the level of detail needed and the type of information resource consulted. Owing to time constraints, health professionals working in patient care require information resources that are readily accessible so information can be

---

[1] It is estimated that there were 4.1 million computers in U.S. classrooms in the 1994–1995 school year, compared with 2.3 million computers in the 1991–1992 school year (GAO, 1996).

efficiently retrieved (e.g., colleagues, textbooks, and handbooks). These health professionals often need a summary of known information (e.g., summary information on a specific chemical and its effects) that accurately answers their questions (Cohen et al., 1982; Northup et al., 1983; Osiobe, 1985). This is particularly true in emergency care situations. When an emergency room physician encounters a patient who has swallowed a household cleaning product, that physician needs immediate factual data (i.e., What product was swallowed? What are the harmful ingredients in that product?) and detailed treatment management information.

On the other hand, health professionals involved in research, policy development, and other activities that allow for a greater length of time to be devoted to locating information will frequently require comprehensive retrieval, an exhaustive search of the literature, and the use of in-depth sources of information (Cohen et al., 1982; Northup et al., 1983; Osiobe, 1985; Wallingford et al., 1990). Additionally, they are most likely not to want summarized information but rather are searching for original data from which they can draw their own conclusions or hypotheses. For example, a study of physicians' use of medical information resources found that 73 percent of the respondents indicated a willingness to spend between 10 and 30 minutes on computer searches (Woolf and Benson, 1989). However, most of these physicians indicated that they would devote this amount of time to conducting research, but not to making clinical decisions. Online databases can serve both the acute and long-term information needs of health professionals if systems are designed with those factors in mind.

Cost

In addition to the expenditure of the health professional's time involved in locating information, the financial costs of online searching may be a factor in determining what sources of information are consulted (Curley et al., 1990; Dalrymple, 1990). Since online charges generally are determined both by the amount of time spent online (the per-minute connect time) and by the amount of information retrieved online, knowledge of the database and expertise in developing relevant search strategies relate directly to the final costs. Because online costs are difficult to determine ahead of time, health professionals may prefer to use CD-ROM resources, which allow for unlimited searching for fixed costs. The extent to which search costs affect searching of online databases as well as potential strategies to simplify pricing structures (thereby providing the user with cost estimates at the outset of the search) need to be considered in future studies.

Accessibility and Relevance

Ease of access and the relevance of the data retrieved are important considerations when seeking information and solutions to diagnostic and treatment problems. The use of textbooks and journals is easy and convenient for most practitioners. Furthermore, discussing a case with a trusted colleague, particularly one who has examined a similar case, may provide more relevance to the situation at hand. These are traditional methods of obtaining information, particularly among many current physicians who were not trained in the use of on-line resources and who are not aware of the benefits of online searching.

Searching a computer database usually results in more extensive amounts of information, if not always completely relevant information,[2] and substantial benefits to the patients. Several studies have demonstrated the benefits of online literature searching to the patients when health professionals have consulted online databases for additional or the most up-to-date information on diagnosis, treatment, and prevention (Haynes et al., 1990; King, 1987; Marshall, 1992). The retrieved information affected patient care in a positive way by changing the way in which a case was handled, contributing to a better-informed clinical decision, or resulting in higher-quality patient care. In 1994, Klein and colleagues examined the association between the use of MEDLINE searches by health care professionals and economic indicators of hospital costs, charges, and length of stay for inpatients. For the patients whose health professionals used MEDLINE and conducted literature searches early in the patients' stay, costs and lengths of stay were significantly lower than those for patients whose health professionals conducted searches later or not at all (Klein et al., 1994; Lindberg et al., 1993).

Health professionals may not be aware of these benefits. Health care trends such as the use of evidence-based medicine approaches that emphasize incorporating the biomedical literature into clinical decision making may provide an impetus for increased searching of bibliographic databases.

CURRENT AND POTENTIAL USERS OF THE TEHIP DATABASES

The following section examines statistics on current TEHIP users and then focuses on potential users, discussing their information needs, the information sources that are currently being consulted for toxicology and environmental health information, barriers to using the TEHIP databases, and opportunities that

---

[2] A study of literature searching by physicians found that physicians deemed 58 percent of the articles retrieved relevant to their initial queries (Gorman et al., 1994).

may increase the use of TEHIP databases. Potential users include all health professionals and other related user communities that might benefit from using the TEHIP databases as a source for toxicology and environmental health information.

Current Users

Toxicology and environmental health information is used by professionals working in a number of fields, including chemical manufacturing, pharmaceutical development, transportation of hazardous materials, environmental law, public advocacy, nursing, and clinical medicine. The general public, as discussed in Chapter 1, is also concerned about the adverse health effects from chemicals and environmental exposures, such as lead, radon, pesticides, smog, dioxin, and carbon monoxide, and may request additional information from their health care professionals or seek out the information themselves.

Only partial statistics on the current use of the NLM databases are available. NLM does have statistics on searches conducted on the NLM servers, however, because NLM licenses its databases for tape, online, and CD-ROM access through commercial vendors, universities, and other institutions (see Chapter 6) complete statistics on all searches of the NLM databases are not available. A recent survey of 2,500 online users of the NLM databases (on the NLM servers) found that 46 percent of the users are health professionals, 20 percent are librarians, and 19 percent are scientists (NLM, 1996). This is in contrast to statistics on use of the TEHIP databases[3] (on the NLM servers), which indicate that the primary users are in industry, with health professionals accounting for only a minor percentage of users. In 1995, 37 percent of the users of HSDB were identified as being from industry, whereas only 6 percent were from the health care community (data supplied by Specialized Information Services Division, NLM). Other TEHIP databases reported similar distributions of users.

MEDLINE is widely used in the health care community as the source for bibliographic citations of the biomedical literature, and training on searching MEDLINE is included in the curricula of many health professional schools. Statistics on searches of the NLM servers show that, as may be expected, the use of MEDLINE far exceeds the use of specialized MEDLARS databases. NLM statistics indicate that in 1995 there were 5,262,329 online searches of MEDLINE, 71,631 of TOXLINE, 30,296 of HSDB, 17,593 of IRIS, 12,793 of

---

[3] These statistics are based on the occupational information provided by users registering for NLM passwords and user IDs.

TRI, 33,427 of DIRLINE, and 13,808 of RTECS (NLM, 1997).[4] MEDLINE contains only limited toxicology and environmental health-related information (including case reports, human toxicity levels, epidemiological studies, and literature reviews).

Potential Users

The committee discussed the health professional communities that potentially have information needs in toxicology and environmental health and use for the information in the TEHIP databases (Box 4.1). The committee took a broad perspective that encompasses a number of groups with interests in environmental health (Chapter 1). The committee realizes that the health professional community does not have homogeneous information needs and that there is wide variation in access to online databases and other information resources. Even with these variations, however, the committee believed that it was worthwhile to discuss potential user communities to provide generalized insights into how NLM might better meet the toxicology and environmental health information needs of these groups. The following list of potential user communities is not meant to be definitive or exhaustive but rather was used by the committee for purposes of discussion:

- primary care professionals (e.g., physicians, nurses, nurse practitioners, and physician assistants) and pharmacists;
- specialists in occupational and environmental health (physicians, nurses, nurse practitioners, physician assistants, industrial hygienists, and safety officers);
- emergency medicine and poison control center personnel (e.g., emergency room health professionals, emergency medical technicians, clinical and medical toxicologists, and specialists in poison information);
- health science librarians and faculty at health professional schools (including medical, nursing, public health, pharmacy, and dental schools);
- environmental health researchers and scientists (including health physicists, epidemiologists, toxicologists, and forensic practitioners);
- patients, the general public, and community organizations (including local emergency planning committees, public librarians, educators, and advocacy and activist organizations); and
- health professionals in local public health departments or in state and federal agencies (e.g., policy advisors, health educators, and public clinic personnel).

---

[4]Note that these statistics only reflect searches on the NLM servers and do not account for searches from commercial or institutional access points.

BOX 4.1
Examples of the Applicability of the TEHIP Databases for the Work of Health Professionals

CLINICALLY RELEVANT INFORMATION

Information that has direct clinical relevance includes the emergency medical treatment information and the human toxicity summaries found in HSDB and TRIFACTS. Additionally, the bibliographic references found in DART provide information on reproductive toxicology, and the TOXLINE/TOXLIT databases provide extensive references on all areas of toxicology and environmental health. Health professionals may find DIRLINE useful in identifying additional information resources and relevant health-related organizations and associations for patients.

RESEARCH INFORMATION

The most likely users of the extensive animal and laboratory data found in several of the TEHIP databases are researchers. CCRIS, RTECS, and GENE-TOX all contain detailed experimental data on carcinogenicity and mutagenicity studies including the dose, target tissue, route of exposure, and test results. RTECS also contains information on skin and eye irritation studies, reproductive studies, and general toxicity studies. EMIC and EMICBACK contain bibliographic references to the literature on mutagenicity studies.

REGULATORY INFORMATION

Health professionals, particularly those working in occupational health settings, can use the information on exposure standards and regulations found in the HSDB, IRIS, RTECS, and TRIFACTS databases. There is extensive information on occupationally permissible exposure levels (OSHA standards and NIOSH recommendations) and the requirements for specific chemicals under a number of environmental laws including CERCLA, TSCA, FIFRA, and the Clean Air and Clean Water Acts.

RISK ASSESSMENT INFORMATION

In working on risk assessments and developing health policies based on risk, health professionals should be aware of the EPA carcinogenic and noncarcinogenic assessments in the IRIS database. This database provides reference doses for oral and inhalation exposures and carcinogenicity assessments based on inhalation or oral exposure.

COMMUNITY ISSUES

The TRI series of databases is a useful information resource for community members and health professionals interested in finding information on industrial emissions in and around their locality. Health professionals should be aware of this information resource as a tool for locating community-based information.

Although the type, depth, and frequency of toxicology and environmental health information needed by each of these groups will differ among individuals within and across the groups, depending on job responsibilities, demographics, training, work or practice setting, time, access, and availability, these groupings provide a framework from which to explore information needs, current strategies for finding information, and potential use of the TEHIP databases. The generalizations presented below draw on input provided to the committee from focus group discussions, responses to the committee questionnaire, and discussions with colleagues and other health professionals. They are presented here as examples from which some general conclusions may be made.

Primary Care Professionals

Information needs. Health care professionals in family health, pediatric health, and women's health are often the first point of contact for patients with environmental health questions and concerns. Primary care professionals often expressed a need for summarized information that they could then provide to patients. For example, pediatricians and pediatric nurse practitioners are frequently asked questions on environmental health exposures by breast-feeding mothers and by mothers concerned about the effects of exposure to hazardous substances during pregnancy. More infrequently but demanding more acute action are cases of poisonings. In these cases primary care professionals need specific and detailed treatment management information immediately.

Information sources. Primary care professionals frequently consult textbooks or colleagues to answer their information needs (Gorman and Helfand, 1995). For acute care situations, primary care professionals call poison control centers for specific treatment management protocols. These resources are rapidly accessible and can provide summarized information. Primary care professionals do not have the time during the patient's visit to search online databases, although searching may be an option prior to providing patient follow-up.

Barriers to using the TEHIP databases. Primary care professionals indicate that the major barriers that they face in searching the TEHIP databases are time, access, training, and the user interface. As indicated above, patient visits are short in duration and require rapidly available information. Primary care professionals may not have immediate access to online databases in the clinic or private-practice office. Additionally, training of primary care professionals has traditionally been based on searching MEDLINE. Therefore, these health professionals are often unfamiliar with the content of the TEHIP databases or the navigation methods necessary to conduct a search.

Opportunities. Trends in the increased availability and use of computers bode well for the use of all online databases as health professionals come to rely more heavily on computer resources for their information. Since primary care professionals are most familiar with searching MEDLINE, any efforts toward standardizing the interfaces (i.e., making MEDLINE and TEHIP database searching comparable from the user interface) will increase the utility of the TEHIP databases to these health professionals. Any modifications or refinements made in the TEHIP databases must be firmly grounded in the realities of daily practice, including issues of need, access, and cost. Training should be focused on those databases of greatest clinical value (e.g., TOXLINE and HSDB). There are many opportunities for training primary care professionals including increased emphasis on occupational and environmental health issues during professional training and continuing education courses.

Occupational and Environmental Health Specialists

Information needs. Occupational and environmental health professionals (e.g., physicians, nurses, nurse practitioners, physician assistants, industrial hygienists, and safety officers) need in-depth information, often on specific exposures. Focus group participants expressed a preference for using bibliographic databases to locate the primary literature rather than relying on summarized information presented in textbooks, factual databases, or other tertiary sources.

Information sources. Specialists noted that they use a number of databases to meet their information needs, including NIOSHTIC, ReproTox, and some of the TEHIP databases, particularly TOXLINE, RTECS, and IRIS.

Barriers to using the TEHIP databases. Specialists often work in academic health care centers, where access to the databases is available; however, they have indicated that they are still often baffled by the user interface to the TEHIP databases (particularly the direct searching interface) and are often not familiar with the range of information available through the TEHIP complement of databases.

Opportunities. Marketing the TEHIP databases to this group of health professionals at occupational and environmental health conferences may be particularly useful. These health professionals are potentially extensive users of the TEHIP databases and indicated that hands-on training by using case studies and real-life scenarios would be useful in helping them to become more comfortable with searching the TEHIP databases. Efforts to improve the user-friendliness of the search interface would also improve the utility of the databases for these health professionals.

Emergency Medicine and Poison Control Center Personnel

Information needs. Health professionals working in poison control centers, in emergency rooms, and on emergency response teams need rapid information on toxic exposures. They are frequently working with exposures to household products and need to be able to specifically translate the brand name of a toxic substance into information that will provide the proper treatment protocol.

Information sources. Specialized information sources have been developed for use by poison control centers. Databases, particularly POISINDEX® are used because they are searchable by brand name and provide links to detailed treatment protocols. Other databases, including the TEHIP databases and reference books, are used by poison control centers to provide additional information.

Although some emergency medical departments have in-house access to the POISINDEX® database, many emergency medical personnel call poison control centers to obtain toxicity information and management recommendations. A 1991 study of emergency physicians in Utah found that 94 percent used poison control centers as their information resource, 78 percent used toxicology textbooks, 34 percent consulted with a colleague, and 24 percent accessed an in-house POISINDEX® database (Caravati and McElwee, 1991).

Barriers to using the TEHIP databases. One of the major barriers to using the TEHIP databases for emergency care is that the TEHIP databases do not have the rapid links between brand names and detailed emergency treatment protocols. Treatment protocols in HSDB are provided from the POISINDEX® database and are made available to HSDB through a reciprocal agreement between NLM and Micromedex, Inc. (see Chapter 2).

Opportunities. There are opportunities for NLM and the American Association of Poison Control Centers to work together on refinements to the TEHIP databases that would make these resources more useful for poison control centers and for the larger health care community.

Health Science Librarians and Faculty at Health Professional Schools

Information needs. Faculty in health professional schools (including nursing, pharmacy, dental, public health, and medical schools) require a working knowledge of the information resources available in toxicology and environmental health to incorporate information resources into case studies and other teaching tools. Librarians working in health science libraries and other academic institutions have a wide range of information needs depending on the type of information requested.

Information sources. Since they work in academic institutions, these health professionals are likely to have access to online and CD-ROM databases. Faculty may rely on textbooks and other reference materials, which are easily accessible in their offices, and on colleagues with specialized expertise, who are also readily available. Health science librarians use numerous online and print information resources.

Barriers to using the TEHIP databases. The complexities of the user interface to the TEHIP databases may deter some faculty members from exploring the scope of the databases. Since numerous databases contain toxicology and environmental health information, faculty members and health science librarians may be unaware of or unfamiliar with the specific databases in the TEHIP complement.

Opportunities. Faculty members would benefit from case studies that have been put together for their use and that include links or references to the TEHIP databases and other information resources in this field. The case studies would be particularly valuable as teaching tools if they were focused on specific topics currently being addressed in the curriculum (e.g., birth defects, cancer, and acute poisonings). Seminars at health professional conferences could focus on demonstrating the scope of these online resources by using case studies. By using the National Network of Libraries of Medicine to increase awareness, not only of these databases but also of the panoply of environmental health and toxicology databases, NLM would efficiently reach health science librarians.

CONCLUSION AND RECOMMENDATION

As seen in the previous section, health professionals have diverse needs for toxicology and environmental health information and face various barriers to fulfilling those information needs. The strengths that NLM can bring to bear on this problem are significant. NLM has the traditional strength of librarians and libraries, which is matching information to the information need. Additionally, NLM has state-of-the-art technical expertise in information organization and retrieval through its extensive research and development program.

The focus groups (Appendix C) and the responses to the questionnaire (Appendix B) sponsored by the committee provide initial insights into the toxicology and environmental health information needs of health professionals and the use of the TEHIP databases by these individuals.[5] Through this input and committee deliberations, the disparate nature of the 16 databases in the TEHIP program became more evident. The committee recognized that the databases are not equally useful for the work of the different segments of the health professional community. Although the committee realized that the current TEHIP complement of databases is the result of both NLM initiatives and interagency agreements and that each database fills an important information niche, the committee believes that the TEHIP program should set priorities that would allow efforts to be focused on those databases that meet the information needs of the greatest number of health professionals. This is particularly critical in light of the fact that the TEHIP program has experienced reduced funding levels from interagency agreements in recent years (Chapter 2).

Comparable to a business marketing strategy that necessitates an understanding of the specific needs of current and potential customers prior to designing and distributing the product, this prioritization of the TEHIP program would first require a more in-depth analysis of the toxicology and environmental health information needs of health professionals. This would include an understanding of the routes of information seeking, the level of detail in the information needed, the types of information required, who searches for information, and the barriers to retrieving information. The goal of this user profile analysis would be to match, as closely as possible, the needs of health professionals with specific TEHIP databases. Upon completion of the user analysis, TEHIP program staff could not only prioritize their training and outreach efforts with an emphasis on those databases that are the most useful to health professionals but could also prioritize the resources that are devoted to the databases with the greatest utility for health professionals.

An in-depth analysis of the user community could be gained from workshops, surveys, and focus groups. Additionally, as will be discussed in Chapter 7, an advisory committee to the TEHIP program composed of potential users and a liaison committee composed of representatives of other federal government agencies involved in environmental health issues would make valuable contributions in assisting NLM in understanding the needs of the user communities.

---

[5]It is important to note the limitations of the focus groups and questionnaire. The committee did not attempt to obtain a random scientific sample for the distribution of the questionnaire. Rather, the questionnaire was distributed both to professional association members and via the Internet. Thus, it was not feasible to determine the rates of response or to characterize the nonresponders. Additionally, responding via the Internet requires some degree of computer expertise. Because travel time and expenses were considerations in inviting focus group participants, most of the participants were drawn from the mid-Atlantic, particularly the Washington, D.C. metropolitan area.

The committee recommends that NLM further expand its efforts to understand the toxicology and environmental health information needs of health professionals and the barriers they face in accessing that information by conducting a detailed user profile analysis. Additionally, the committee recommends that the results from that analysis be used to set priorities for subsequent efforts of the TEHIP program.

REFERENCES

Blackwelder MB, Dimitroff A. 1996. The image of health science librarians: How we see ourselves and how patrons see us. *Bulletin of the Medical Library Association* 84(3):345–350.

Caravati EM, McElwee NE. 1991. Use of clinical toxicology resources by emergency physicians and its impact on poison control centers. *Annals of Emergency Medicine* 20(2):147–150.

Cohen SJ, Weinberger M, Mazzuca SA, McDonald CJ. 1982. Perceived influence of different information sources on the decision-making of internal medicine housestaff and faculty. *Social Science and Medicine* 16(14):1361–1364.

Connelly DP, Rich EC, Curley SP, Kelly JT. 1990. Knowledge resource preferences of family physicians. *Journal of Family Practice* 30(3):353–359.

Corcoran-Perry S, Graves J. 1990. Supplemental-information-seeking behavior of cardiovascular nurses. *Research in Nursing and Health* 13:119–127.

Covell DG, Uman GC, Manning PR. 1985. Information needs in office practice: Are they being met? *Annals of Internal Medicine* 103:596–599.

Curley SP, Connelly DP, Rich EC. 1990. Physicians' use of medical knowledge resources: Preliminary theoretical framework and findings. *Medical Decision Making* 10:231–241.

Dalrymple PW. 1990. CD-ROM MEDLINE use and users: Information transfer in the clinical setting. *Bulletin of the Medical Library Association* 78(3):224–232.

GAO (General Accounting Office). 1996. *Consumer Health Information: Emerging Issues*. Publication No. GAO/AIMD-96-86. Washington, DC: U.S. General Accounting Office.

Gorman PN, Ash J, Wykoff L. 1994. Can primary care physicians' questions be answered using the medical literature? *Bulletin of the Medical Library Association* 82:140–146.

Gorman PN, Helfand M. 1995. Information seeking in primary care: How physicians choose which clinical questions to pursue and which to leave unanswered. *Medical Decision Making* 15:113–119.

Gruppen LD. 1990. Physician information seeking: Improving relevance through research. *Bulletin of the Medical Library Association* 78(2):165–172.

Haynes RB, McKibbon A, Walker CJ, Ryan N, Fitzgerald D, Ramsden MF. 1990. On-line access to MEDLINE in clinical settings: A study of use and usefulness. *Annals of Internal Medicine* 112:78–84.

Huth EJ. 1985. Needed: An economics approach to systems for medical information. *Annals of Internal Medicine* 103(4):617–619.

IOM (Institute of Medicine). 1988. *Role of the Primary Care Physician in Occupational and Environmental Medicine.* Washington, DC: National Academy Press.

King DN. 1987. The contribution of hospital library information services to clinical care: A study in eight hospitals. *Bulletin of the Medical Library Association* 75(4):291–301.

Klein MS, Ross FV, Adams DL, Gilbert CM. 1994. Effect of online literature searching on length of stay and patient care costs. *Academic Medicine* 69:489–495.

Lindberg DAB, Siegel ER, Rapp BA, Wallingford KT, Wilson SR. 1993. Use of MEDLINE by physicians for clinical problem solving. *Journal of the American Medical Association* 269:3124–3129.

Lockyer JM, Parboosingh JT, McDougal GM, Chugh U. 1985. How physicians integrate advances into clinical practice. *Mobius* 5(2):5–12.

Maizell RE. 1967. Continuing education in technical information services. *Journal of Chemical Documentation* 7:115.

Marshall JG. 1992. The impact of the hospital library on clinical decision making: The Rochester study. *Bulletin of the Medical Library Association* 80:162–178.

NLM (National Library of Medicine). 1995. *National Library of Medicine: Programs and Services, Fiscal Year 1994.* Bethesda, MD: NLM.

NLM. 1996. *Survey of Online Customers: Usage Patterns and Internet Readiness.* NIH Publication No. 96-4181. Bethesda, MD: NLM.

NLM. 1997. *National Library of Medicine: Programs and Services, Fiscal Year 1995.* Bethesda, MD: NLM. NIH Publication No. 97-256.

Northup DE, Moore-West M, Skipper B, Teaf SR. 1983. Characteristics of clinical information-seeking: Investigation using critical incident technique. *Journal of Medical Education* 58:873–881.

Osheroff JA, Forsythe DE, Buchanan BG, Bankowitz RA, Blumenfeld BH, Miller RA. 1991. Physician's information needs: Analysis of questions posed during clinical teaching. *Annals of Internal Medicine* 114:576–581.

Osiobe SA. 1985. Use of information resources by health professionals: A review of the literature. *Social Science and Medicine* 21(9):965–973.

Wallingford KT, Humphreys BL, Selinger NE, Siegel ER. 1990. Bibliographic retrieval: A survey of individual users of MEDLINE. *MD Computing* 7(3):166–171.

Williams RM, Baker LM, Marshall JG. 1992. *Information Searching in Health Care.* Thorofare, NJ: SLACK Inc.

Williamson GW, German PS, Weiss R, Skinner EA, Bowes F. 1989. Health science information management and continuing education of physicians. *Annals of Internal Medicine* 110:151–160.

Woolf SH, Benson DA. 1989. The medical information needs of internists and pediatricians at an academic medical center. *Bulletin of the Medical Library Association* 77(4):372–380.

5

Increasing Awareness: Training and Outreach

An important component of expanding the use of toxicology and environmental health information resources is increasing the awareness of these resources in the health professional and other interested user communities. Potential users of online toxicology and environmental health databases must be cognizant of the existence of the databases and of their content, must be computer literate (assuming that the user will perform his or her own search), and must have some familiarity with toxicology and environmental health data in order to interpret the retrieved information correctly.

Government agencies traditionally are not involved in marketing their products or databases. However, in 1987 an amendment to the NLM Act added to NLM's mission the function of publicizing the availability of NLM's products and services (Public Law 100-202, section 215). Furthermore, in 1988 the Senate Committee on Appropriations recognized that

> The nation's immense investment in biomedical research can be maximized only if there are efficient channels for disseminating research results, and these the library [NLM] provides through its computerized MEDLARS services and the regional medical library network. The Committee believes that this program should be expanded to reach all American health professionals, wherever located, so that they will be able to take advantage of the library's information services (U.S. Senate Committee on Appropriations, 1988, p. 145).

This direction toward increased dissemination has led to the expansion of NLM's outreach and training programs, which are discussed throughout this chapter. Additionally, this chapter makes recommendations aimed at increasing the awareness and utility of not only the TEHIP databases but also other toxicology and environmental health information resources.

TRAINING

For health professionals to efficiently search toxicology and environmental health databases and effectively use the information in those resources, there are several educational requirements. Health professionals need a working knowledge of computers, especially online searching skills, and an understanding of the strengths and weaknesses of the information available in this field.

Training in the Use of Computers and Online Database Searching

As discussed in Chapter 4, computer use is largely a matter of demographics. Younger health professionals are more likely to feel comfortable with computer use and to have become accustomed to retrieving information through computers. There is, however, a continuing need to train health professionals about specific databases and the use of health-related information resources.

Koschmann (1995) categorized computer training into three types: (1) learning about computers, (2) learning with computers (i.e., computer-assisted instruction and the use of computers as specialized tools for instruction), and (3) learning through computers (i.e., incorporating computers into student's work and assignments on a daily basis). The author concluded that although there is a place for each of these in the educational process, it is learning through computers that is most effective in preparing for lifelong learning.

A 1984 report by the Association of American Medical Colleges recommended the introduction of computer training into medical education (AAMC, 1984; Matheson and Lindberg, 1984). Although most medical schools now teach basic computer skills (e.g., literature searching and word processing), a survey of 1992 medical school graduates found that 39 percent believed that the computer training that they had received was inadequate (Anderson, 1993; Hersh, 1992). Medical, nursing, and other health professional schools use a number of computer-assisted instruction programs for tutorials and computer simulations of clinical decision points (Hoffer and Barnett, 1990). Additionally, recognizing the vastness of the biomedical knowledge base and the necessity of lifelong learning, many health professional schools are using approaches, such as problem-based learning, that focus on independent learning and that incorporate learning through computers by emphasizing the frequent use of information resources, such as online databases, to solve clinical problems (Earl et al., 1996; Rankin, 1992; Schilling et al., 1995). Approaches to computer training vary between institutions. Some medical schools require an informatics or literature-searching practicum, whereas in many medical schools this training is available only through noncredit classes (Florance et al., 1995; Ikeda and Schwartz, 1992; Pao et al., 1993). Responsibilities for computer training are often shared between departments of medical informatics (where available) and the health sciences library.

Online database searching, particularly MEDLINE searching, is often incorporated into continuing education courses, and evidence-based medicine continues this emphasis by using the biomedical literature to inform clinical decision making throughout the physician's career.

Training in Occupational and Environmental Health

Answering questions and making proper diagnoses of exposure-related illnesses pose challenges for any health professional. As discussed in Chapter 1, the complexities of interpreting toxicology and environmental health data are substantial. Although no health professional can be expected to know the toxic effects of all chemicals, it is critical that health professionals be informed about the issues and familiar enough with the field to consider environmental and occupational exposures in assessing a patient's symptoms, making a diagnosis, answering a patient's questions, and counseling patients about environmental health risks.

However, health professionals, especially clinicians, receive limited education and training in toxicology and environmental health in part because of an overcrowded and increasingly specialized curriculum (Graber et al., 1995; Snyder et al., 1994). One study concluded that medical students receive fewer than 6 hours of occupational and environmental medicine during 4 years of medical school (Burstein and Levy, 1994). A survey of 89 departments of internal medicine found that 51 (57 percent) did not offer programs or clinics in occupational and environmental medicine and that only 20 programs (22 percent) offered clinical occupational medicine experience to residents, in most cases this training was elective (Cullen, 1987). A survey of 423 accredited baccalaureate schools of nursing found that while most of those responding included occupational and environmental health content in their curricula, fewer than half offered course content in epidemiology, toxicology, industrial hygiene, and occupational health nursing concepts and practice—related fields of knowledge essential to the understanding and management of such problems (Rogers, 1991). A 1994 study of graduate nursing programs in public health or community nursing found that only 17 percent required a course in environmental health (Ostwalt and Josten, 1994). Several recent IOM studies have focused on strategies for enhancing the environmental health content in health sciences curriculum and continuing education courses (IOM, 1988, 1995a,b), and this committee supports the recommendations of those reports (Box 5.1).

> **BOX 5.1**
> **Previous IOM Recommendations on the Training of
> Health Professionals in Occupational and Environmental Health**
>
> *Nursing, Health, and the Environment* (IOM, 1995b)
> • Environmental health concepts should be incorporated into all levels of nursing education.
> • Environmental health content should be included in nursing licensure and certification examinations.
> • Expertise in various environmental health disciplines should be included in the education of nurses.
> • Environmental health content should be an integral part of lifelong learning and continuing education for nurses.
> • Professional associations, public agencies, and private organizations should provide more resources and educational opportunities to enhance environmental health in nursing practice.
>
> *Environmental Medicine* (IOM, 1995a)
> Graduating medical students should:
> • understand the influence of environment and environmental agents on human health based on knowledge of relevant epidemiologic, toxicologic, and exposure factors;
> • be able to recognize the signs, symptoms, diseases, and sources of exposure relating to common environmental agents and conditions;
> • be able to elicit an appropriately detailed environmental exposure history, including a work history, from all patients;
> • be able to identify and access the informational, clinical, and other resources available to help address patient and community environmental health problems and concerns;
> • be able to discuss environmental risks with their patients and provide understandable information about risk-reduction strategies in ways that exhibit sensitivity to patients' health beliefs and concerns; and
> • be able to understand the ethical and legal responsibilities of seeing patients with environmental and occupational health problems or concerns.
>
> *Role of the Primary Care Physician in Occupational and Environmental Medicine* (IOM, 1988)
> • There should be a better representation of occupational and environmental medicine in the medical school curriculum.
> • If occupational and environmental medicine are to prosper in academia, a vigorous research program is required.
> • Residency programs directed toward the production of general physicians in both internal medicine and family practice should be adjusted to provide more active clinical experience in occupational and environmental medicine.

Current efforts to increase training in environmental health include the addition of occupational and environmental health questions on certification examinations, the awarding of curriculum grants for environmental health training by the National Institute of Environmental Health Sciences and by the Agency for Toxic Substances and Disease Registry, and the establishment of task forces by the Children's Environmental Health Network to develop training materials and training guides for pediatric residency faculty and other health professionals.

Training in Health Information Management

There is a continuing need for individuals cross-trained in computers and telecommunications, information sciences, and biomedicine. Individuals specializing in medical informatics use basic research in cognitive science, decision science, logic, statistics, computer science, and a number of other fields to develop and deploy applications that are specific to the needs of health professionals (Greenes and Shortliffe, 1990). Additionally, they are involved in planning and policy development and work to integrate information technology in health care and medical education. One of the emerging areas in this field is public health informatics, which will incorporate environmental health information issues.

NLM offers institutional training grants and individual fellowships to promote interdisciplinary training in medical informatics. Of those individuals completing the training programs, it is estimated that most are working in research environments, with approximately two-thirds in academic medicine and one-third in health-related research and development (Wallingford et al., 1996).

OUTREACH

To assist in strategic planning on outreach issues, the NLM Board of Regents convened an Outreach Planning Panel in 1988 that was mandated to identify opportunities for improving the access to and dissemination of NLM's information resources (NLM, 1989). The initiatives proposed by that panel have been implemented into what is now an extensive outreach effort. NLM has established an Office of Outreach Development that works with its interdivisional Outreach Coordinating Committee to plan, develop, and evaluate NLM's outreach program. Recently, NLM conducted a 5-year review of its outreach efforts, which noted the expansion from 16 outreach projects in 1989 to approximately 300 projects (involving more than 500 libraries and other institutions) in 1994 (Wallingford et al., 1996).

Current NLM-Wide Outreach Activities

National Network of Libraries of Medicine

NLM's outreach efforts are primarily conducted through the National Network of Libraries of Medicine (NN/LM), a nationwide network of more than 4,500 local medical libraries (primarily hospital libraries), more than 140 Resource Libraries (primarily at medical schools), and eight Regional Medical Libraries (RMLs) (Figure 5.1). The medical librarians and information specialists at each NN/LM member library play an integral role in educating, training, and providing access to NLM's resources.

Within each geographic region, the RMLs manage outreach activities that focus on connecting unaffiliated, rural, and minority health professionals with their local medical libraries and with NLM products and services. NLM supports Grateful Med demonstration and training projects through NN/LM, and since 1994, RMLs have awarded American Medical Association Category 1 Continuing Medical Education credits to those attending Grateful Med training sessions. Through the National Online Training Center sponsored by NN/LM, health professionals can register online for classes that are held regionally.

The current NLM contracts with the eight RMLs emphasize public health outreach programs. Each RML develops projects targeted for populations in its geographic region, and RML-sponsored outreach activities have been developed to reach a range of health professionals including physicians, nurses, dentists, pharmacists, therapists, and health administrators (Wallingford et al., 1996).

Extramural Grants

One of NLM's major outreach initiatives is a program that provides extramural grants to hospital libraries and other health science libraries for the purposes of introducing and expanding computer and telecommunications technologies. Four grant mechanisms have significance to outreach efforts:

- Information Access Grants provide assistance, primarily to small hospital libraries, for introducing computer technology.
- Information Systems Grants (ISGs) provide funds for improving or expanding access to information technologies and are primarily targeted to larger hospitals and academic health centers (e.g., an ISG provided funds for the establishment of a statewide network in Alaska that links rural hospitals and the medical library at the University of Alaska in Anchorage [Wallingford et al., 1996]).
- The Integrated Advanced Information Management Systems (IAIMS) initiative is designed to encourage the integration of multiple computerized

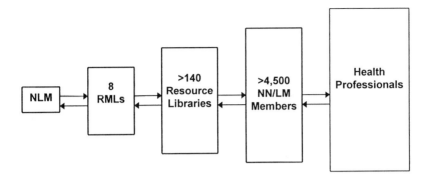

FIGURE 5.1 National Network of Libraries of Medicine. SOURCE: Adapted from NLM (1986).

information systems located throughout a hospital or medical center. IAIMS grants provide institutions with funding to plan and develop institution-wide computer networks that may include clinical, research, and library resources; administrative systems; and other computer functions such as word processing and email (Broering, 1992; Fuller, 1992; NLM, 1989; Roderer and Clayton, 1992).

- Internet Connection Grants are available through NLM. They enable health science libraries and other organizations to initiate Internet connections or to extend existing connections to outlying institutions (Corn and Johnson, 1994).

Training in Database Searching

NLM offers a workshop on searching ChemID, CHEMLINE, TOXLINE, TOXLIT, and the TOXNET databases. These 2-day workshops are free-of-charge and are frequently held in conjunction with MEDLINE training courses. The TOXNET workshop is conducted by Specialized Information Services Division (SIS) staff and is held at the National Institutes of Health (NIH) campus in Bethesda, Maryland. The workshop features hands-on searching of the databases and provides extensive documentation and sample search printouts for the students to take back to their workplace. This class is a specialized module in NLM's entire training program that features introductory and advanced classes on MEDLARS searching.

Although these courses are available, there is only limited attendance. In FY 1995, 14 classes were offered on the ChemID, CHEMLINE, TOXLINE, TOXLIT, TOXNET databases, and of those, 6 classes were canceled. Sixty individuals received training in the 8 classes that were held. The reasons for the limited attendance may include the length of the course, location, limited publicity, or lack of awareness of the utility of these databases.

NLM is developing a microcomputer-based toxicology courseware program that will provide an introduction to the fields of toxicology and environmental health for undergraduate students (NLM, 1995). It would be beneficial for this courseware, when it is released, to go beyond a stand-alone program and to be available on the World Wide Web as a case-based learning resource.

Current TEHIP-Specific Outreach Activities

In addition to working with NLM-wide outreach activities, SIS has been active in targeting outreach activities to those populations particularly interested in environmental health information. Authoritative toxicology and environmental health information is of particular importance to underserved populations, those communities and locales where medical care is not readily or thoroughly available. These are often the same low-income or minority communities that are disproportionately exposed to hazardous environmental conditions at work and in the home (Averill and Samuels, 1992; Benson, 1995). One of the objectives set forth by the U.S. Department of Health and Human Services' Subcommittee on Environmental Justice, which is examining strategies to address environmental injustice issues and the associated adverse human health and environmental effects, is "to make environmental and occupational health data more available to the public" (DHHS Subcommittee on Environmental Justice, 1995). SIS sponsors several outreach programs that provide information to the general public, to underserved communities and the health professionals who serve them, and to other health professionals who are not affiliated with a health sciences library.

Toxicology Information Outreach Project

In 1991, SIS implemented a pilot Toxicology Information Outreach Project with the objective of strengthening the capacity of Historically Black Colleges and Universities (HBCUs) to train medical and other health professionals in the use of NLM's toxicology and environmental health information resources. Nine HBCUs currently participate in the pilot program: Drew University School of Medicine and Science, Florida A&M University, Howard University, Meharry Medical College, Morehouse School of Medicine, Texas Southern University, Tuskegee University, University of Arkansas at Pine Bluff, and Xavier University. The program provides a workstation with computer-based tutorials, Grateful Med software, instructional materials, and free online access to the NLM databases. Training is offered at each institution to researchers, instructors, and students, and in some cases health professionals in neighboring communities have attended training sessions.

Additionally, NLM collaborates with the Agency for Toxic Substances and Disease Registry (ATSDR) to train members of other HBCUs and minority institutions that offer environmental and occupational health programs. Three-day training classes have been held at the Oak Ridge Institute for Science and Education and at Howard University and have been attended by members of 44 additional HBCUs and minority institutions.

The Toxicology Information Outreach Project has had a significant impact on the curriculum and on outreach programs (Box 5.2). Use of the NLM databases has been incorporated into a number of courses at the undergraduate and graduate levels. Additionally, several universities have initiated their own outreach efforts involving educating junior high and high school students, college undergraduates, and local health professionals in accessing and searching NLM's toxicology and environmental health databases (Wallingford et al., 1996).

BOX 5.2
Howard University

Howard University has been a leader in implementing NLM's toxicology and environmental health information resources into its medical school and undergraduate curriculum. The initial workstation supplied through the NLM Toxicology Information Outreach Project has been supplemented by six additional workstations in the Department of Pharmacology Medical Informatics Lab, which connects students to NLM resources and to a number of other online toxicology databases. Interest in computer databases has led to the development of medical informatics courses taught through the College of Medicine and the Graduate School of Arts and Sciences. Additionally, use of the NLM databases has been incorporated into the course requirements for a number of graduate-level health sciences departments including pharmacology and biochemistry.

Since 1993, Howard University has hosted an annual training class on the NLM toxicology and environmental health information resources that is jointly sponsored by ATSDR, NLM, and the Environmental Justice Office of the EPA. The course is offered to members of HBCUs in the Lower Mississippi Delta and is a part of the Mississippi Delta Project.

Wheaton Regional Library Project

SIS is actively involved in a collaborative project between NLM and the Health Information Center (HIC) of the Wheaton Regional Library of the Montgomery County (Maryland) Department of Public Libraries. This project promotes access to information resources on environmental health and HIV/AIDS. NLM supplies free access to the relevant online databases on these topics (including AIDSLINE, AIDTRIALS, HSDB, and TRI), in addition to other services, such as developing presentations for community groups; providing books, periodicals, and other documents; and providing assistance in linking the HIC World Wide Web homepage with other Internet sites on environmental health and HIV/AIDS. Through this collaborative effort, environmental health information is being disseminated to several interested user communities including the general public and health professionals (such as school nurses) who are not affiliated with a health sciences library.

Exhibits

NLM and NN/LM staff exhibit the MEDLARS databases and new NLM products and services at a number of national, regional, and local health professional meetings and conferences. Additionally, SIS and NN/LM staff schedule a number of exhibits specifically to demonstrate the TEHIP databases at relevant national and regional meetings and conferences for health professionals. In FY 1996, the TEHIP program exhibit schedule included the annual conferences of the American Industrial Hygiene Association, the National Environmental Health Association, and the Society of Toxicology and the American Occupational Health Conference. To ensure that TEHIP program resources and databases are featured at all relevant regional and local meetings and conferences, communication and collaboration is essential between SIS and NN/LM staff and health professional organizations at the national, state, and local levels.

FUTURE DIRECTIONS AND RECOMMENDATION

The committee believes that the user profile analysis recommended in Chapter 4 will be particularly helpful in outreach and training efforts. That analysis will allow NLM to set priorities and will provide information on the databases of greatest use to specific groups of health professionals. Thus, outreach and training efforts can be targeted so that resources are used most efficiently. As will be discussed in Chapter 7, an evaluation component, incorporated into the initial planning stages for each project, is critical in assessing the impact of training and outreach efforts.

Emphasizing the Broad Spectrum of Information Resources

As discussed in Chapter 3, the committee believes that the TEHIP program can make significant contributions to health professionals by providing information on the broad spectrum of databases and other information resources available in toxicology and environmental health, including but not limited to the TEHIP databases. By using the traditional expertise of libraries in providing reference services, outreach and training efforts could provide health professionals with information on the scope of relevant information resources and their available access points.

Targeting Outreach and Training Efforts

Outreach and training efforts are most effective when they are focused on the specific interests of the audience. A number of methods of incorporating this approach are available. Careful examination of the user analysis (see Chapter 4) could assist NLM in focusing its outreach and training efforts on those databases that best meet the needs of the user community. Focus group participants and the committee members believe that case studies are particularly effective since they can incorporate the use of relevant toxicology and environmental health information resources into clinical situations dealing with a particular topic (e.g., birth defects and cancer) of interest to the audience. Similarly-focused efforts can be used in continuing education programs and would be effective in the growing use of the World Wide Web to provide continuing education courses.

NLM and NN/LM staff currently exhibit the MEDLARS databases at a number of health professional conferences. Focus group participants indicated that they would be interested in having NLM conduct workshops or do teleconferences using search scenarios relevant to the topics at specific conference (e.g., environmental exposures of children at pediatrics conferences). These workshops would be ideal places to provide health professionals with information on the wide range of relevant toxicology and environmental health information resources.

Although most health professionals do not require intense specialized training on the TEHIP databases, such as the training currently provided through NLM's National Online Training Program, this training is useful for certain individuals including librarians and scientists who may be performing in-depth searches of the TEHIP databases. The committee believes that the key to effective outreach and training is to focus those efforts by demonstrating the utility of the TEHIP databases and other similar information resources to health professionals.

Use Existing Dissemination Mechanisms

To reach the wide range of health professional user communities who can benefit from toxicology and environmental health information, it is important to use already-existing dissemination networks and current environmental health efforts. Committee members are aware of a number of dissemination mechanisms, including those described below, and encourage NLM to use these and other methods. Efforts should focus on those information resources that will be most useful for the targeted audience.

Information is currently disseminated by EPA to local emergency planning committees which have a strong interest in environmental health issues. Tying into that network to disseminate information about relevant TEHIP and other information resources would be advantageous. Environmental justice efforts are another focal point for outreach at the regional and community levels. Additionally, the NN/LM is an extensive existing network that is readily available. Since a large focus of NN/LM's outreach efforts is on Grateful Med training, incorporating searches involving the TEHIP databases into Grateful Med search examples would be a low-cost yet effective means of training. As seen in SIS's current cooperative project with the Wheaton Regional Public Library, public libraries are very effective in reaching many groups interested in environmental health issues, including patients, the general public, and health professionals such as school nurses who are not tied into other library networks. Local public health departments and poison control centers are also resources that would benefit from close ties with the TEHIP program. Publishing articles or informational advertisements in health professional publications is another way of reaching a large audience; these resources include scientific journals, and professional society newsletters and Web sites. These are only a few examples of networks and activities that are in place and that can be used to inform health professionals about toxicology and environmental health information resources that are relevant to their work.

> The committee recommends that NLM's training and outreach efforts in toxicology and environmental health information be increased to improve awareness and recognition of these resources. Mechanisms that may improve awareness include:
>
> - emphasizing the broad spectrum of toxicology and environmental health information resources,
> - matching training to meet the specific needs of the target audiences, and
> - expanding the use of already-existing distribution mechanisms for promoting the availability of toxicology and environmental health information.

REFERENCES

AAMC (Association of American Medical Colleges). 1984. Physicians for the Twenty-First Century: Report of the Project Panel on the General Professional Education of the Physician and College Preparation for Medicine. *Journal of Medical Education* 59(Part 2).

Anderson MB. 1993. Medical education in the United States and Canada revisited. *Academic Medicine* 68(6 Suppl):S55–S63.

Averill E, Samuels SW. 1992. International occupational and environmental health. In: Rom WN, ed. *Environmental and Occupational Medicine*. Boston: Little, Brown, and Company. Pp. 1357–1364.

Benson L. 1995. Environmental health in minority communities. In: Brooks SM, Gochfeld M, Jackson RJ, Herzstein J, Schenker MB, eds. *Environmental Medicine*. St. Louis: Mosby. Pp. 412–418

Broering NC. 1992. Fulfilling the promise: Implementing IAIMS at Georgetown University. *Medical Progress Through Technology* 18(3):137–149.

Burstein JM, Levy BS. 1994. The teaching of occupational health in U.S. medical schools: Little improvement in 9 years. *American Journal of Public Health* 84(5):846–849.

Corn M, Johnson FE. 1994. Connecting the health sciences community to the Internet: The NLM/NSF grant program. *Bulletin of the Medical Library Association* 82(4):392–395.

Cullen M. 1987. *The Challenge of Teaching Occupational and Environmental Medicine in Internal Medicine Residencies*. Commissioned paper prepared for the Institute of Medicine. July.

DHHS (U.S. Department of Health and Human Services), Subcommittee on Environmental Justice. 1995. Strategic elements for environmental justice. *Environmental Health Perspectives* 103(9):796–800.

Earl MF, Hensley K, Fisher JS, Kelley MJ, Merrick D. 1996. Faculty involvement in problem-based library orientation for first-year medical students. *Bulletin of the Medical Library Association* 84(3):411–413.

Florance V, Braude RM, Frisse ME, Fuller S. 1995. Educating physicians to use the digital library. *Academic Medicine* 70(7):597–602.

Fuller SS. 1992. Creating the future: IAIMS planning premises at the University of Washington. *Bulletin of the Medical Library Association* 80(3):288–293.

Graber DR, Musham C, Bellack JP, Holmes D. 1995. Environmental health in medical school curricula: Views of academic deans. *Journal of Occupational and Environmental Medicine* 37(7):807–811.

Greenes RA, Shortliffe EH. 1990. Medical informatics: An emerging academic discipline and institutional priority. *Journal of the American Medical Association* 263(8): 1114–1120.

Hersh WR. 1992. Informatics: Development and evaluation of information technology in medicine. *Journal of the American Medical Association* 267(1):167–168.

Hoffer EP, Barnett GO. 1990. Computers in medical education. In: Shortliffe EH, Perreault LE, eds. *Medical Informatics: Computer Applications in Health Care*. Reading, MA: Addison-Wesley.

Ikeda NR, Schwartz DG. 1992. Impact of end-user search training on pharmacy students: A four year follow-up study. *Bulletin of the Medical Library Association* 80:124–130.

IOM (Institute of Medicine). 1988. *Role of the Primary Care Physician in Occupational and Environmental Medicine.* Washington, DC: National Academy Press.

IOM. 1995a. *Environmental Medicine: Integrating a Missing Element into Medical Education.* Washington, DC: National Academy Press.

IOM. 1995b. *Nursing, Health, and the Environment: Strengthening the Relationship to Improve the Public's Health.* Washington, DC: National Academy Press.

Koschmann T. 1995. Medical education and computer literacy: Learning about, through, and with computers. *Academic Medicine* 70(9):818–821.

Matheson N, Lindberg D. 1984. Subgroup report on medical information science skills. *Journal of Medical Education* 59(Part 2):155–159.

NLM (National Library of Medicine). 1989. *Improving Health Professionals' Access to Information. Report of the Board of Regents.* Bethesda, MD: NLM.

NLM. 1995. *National Library of Medicine: Programs and Services, Fiscal Year 1994.* Bethesda, MD: NLM.

Ostwalt S, Josten L. 1994. *Preparation of Public Health Nurses for Leadership Positions.* Presentation at the Association of Community Health Nursing Educators Conference. San Antonio, TX. June.

Pao ML, Grefsheim SF, Barclay ML, Woolliscroft JO, McQuillan M, Shipman BL. 1993. Factors affecting students' use of MEDLINE. *Computers and Biomedical Research* 26:541–555.

Rankin JA. 1992. Problem-based medical education: Effect on library use. *Bulletin of the Medical Library Association* 80(1):36–43.

Roderer NK, Clayton PD. 1992. IAIMS at Columbia Presbyterian Medical Center: Accomplishments and challenges. *Bulletin of the Medical Library Association* 80(3):253–262.

Rogers B. 1991. Occupational health nursing education: Curricular content in baccalaureate programs. *American Association of Occupational Health Nurses Journal* 39(3):101–108.

Schilling K, Ginn DS, Mickelson P, Roth LH. 1995. Integration of information-seeking skills and activities into a problem-based curriculum. *Bulletin of the Medical Library Association* 83(2):176–183.

Snyder M, Ruth V, Sattler B, Strasser J. 1994. Environmental and occupational health education: A survey of community health nurses' need for educational programs. *American Association of Occupational Health Nurses Journal* 42(7):325–328.

U.S. Senate Committee on Appropriations. 1988. *Departments of Labor, Health and Human Services, and Education and Related Agencies Appropriation Bill, 1989: Report to Accompany H.R. 4783.* 100th Cong., 2d sess. S. Rept. 100-399. P. 145.

Wallingford KT, Ruffin AB, Ginter KA, Spann ML, Johnson FE, Dutcher GA, Mehnert R, Nash DL, Bridgers JW, Lyon BJ, Siegel ER, Roderer NK. 1996. Outreach activities of the National Library of Medicine: A five-year review. *Bulletin of the Medical Library Association* 84(2 Suppl).

6

Accessing and Navigating the TEHIP Databases

Although the audience served by the TEHIP program encompasses a number of different user communities with a variety of information needs and a range of sophistication in searching online databases, there are common issues involved in accessing and navigating the TEHIP databases. These issues encompass the entire search process from establishing the necessary computer connections, to the development of a search strategy, to the evaluation of the search results. Recommendations to NLM involving the refinements in the technology associated with the TEHIP databases will facilitate use by all interested searchers. This chapter focuses on searching the TEHIP databases directly through NLM and does not address the diverse interfaces and search mechanisms available through commercial vendors or academic institutions. Several search options for NLM users are discussed throughout the chapter and are highlighted in Box 6.1. Additionally, this chapter discusses a range of issues that affect access and navigation and presents the committee's short- and long-term recommendations for refining the TEHIP databases.

ACCESS: GETTING CONNECTED TO THE DATABASES

Although this is a time of increased computer use by health professionals, input to the committee from focus group participants indicates that access to the TEHIP database system is not readily available (Appendix C). Access to the databases was selected as the primary factor that limits use of the TEHIP databases by those individuals responding to a committee questionnaire (Appendix B). Although access to MEDLINE is a standard feature of end-user

> **BOX 6.1**
> **Methods of Searching the TEHIP Databases**
>
> **Grateful Med:** Grateful Med was developed to simplify MEDLINE searching. However, input screens have gradually been added for other databases, and users can now access 20 NLM databases through Grateful Med. Grateful Med currently offers input screens for all of the TEHIP databases except DART/ETICBACK, EMIC/EMICBACK, GENE-TOX, and TRIFACTS. In 1996 Internet Grateful Med was introduced and is currently available for searching MEDLINE, HealthSTAR, AIDSLINE, and PREMEDLINE.
>
> **Direct Searching:** The only fully operational method available for accessing all of the TEHIP databases is direct searching using NLM's search command language. Direct searching involves the use of commands, mnemonics (abbreviations indicating database fields), and Boolean logic for combining terms. This interface requires a high level of search proficiency and expertise.
>
> **Menu Searching:** Two of the TEHIP databases, TRI and HSDB, have menu search interfaces with step-by-step options for assisting users in developing a search strategy.
>
> **Experimental World Wide Web Interface:** SIS staff are in the process of developing a new approach to searching the TEHIP databases that uses the hypertext and graphics capabilities available through the World Wide Web.

workstations in health science libraries, fewer library workstations provide end-user access to the TEHIP databases.

NLM is in the somewhat unique position of playing a number of different roles in the database process. For a few of the TEHIP databases, such as HSDB and ChemID, NLM is the creator of the database and is responsible for the content. In other cases, such as CCRIS and IRIS, other federal agencies are responsible for content and NLM is responsible for maintaining the database on the NLM computer servers (see Chapter 2). These agencies may provide additional access points for their databases such as leasing them in online or CD-ROM formats. For all of the databases in the MEDLARS system, NLM acts as a database provider and offers online search access for registered users. Additionally, NLM as a database producer leases its databases to commercial vendors, academic institutions, libraries, and other organizations. The result is a variety of access routes available to users interested in searching the NLM databases. However, the number of access routes varies depending on the specific database (Figure 6.1). MEDLINE is offered by many commercial vendors in numerous CD-ROM and online formats, and most health science libraries and biomedical organizations provide their users with access to MEDLINE. On the other hand, the more specialized databases have fewer points of access, and there is only one access point—direct searching—that allows the user to connect with the entire TEHIP complement of all 16 databases. This is particularly problematic because direct searching's complex command-line interface requires the most expertise in online searching (see discussion below).

There are advantages and disadvantages to the various access modes. Commercial vendors may add value to the databases by providing easy-to-use search interfaces and may integrate the database into the vendor's network of databases on similar topics. Academic institutions and hospitals may offer access to the NLM databases as a component of the campus-wide network with connections to a wide range of applications including library services and email. Accessing the databases directly through NLM is lower in cost than accessing them through commercial databases because the search costs (user fees) are charged only to cover access costs (e.g., telecommunications; see Chapter 2). NLM's federal budget appropriation is used to build and maintain the databases.

As with many databases, users are required to register to access the NLM databases and are assigned an NLM User Identification (ID) code and password.[1] In June 1996 a new feature was added to Internet Grateful Med (Box 6.1) allowing users to establish a new NLM account online and to use it for immediate searching. In most cases, however, registration is handled through mail or fax. There is no fee to register for a user ID code, and the code allows access to all of the NLM databases. However, the time and effort required to apply for the ID and the inconvenience of having another set of numbers and codes to remember may present barriers to some users. Simplifying the registration process and promoting the availability of online registration will encourage health professionals and other interested individuals[2] to become connected to the TEHIP databases.

NLM offers technical support to individuals having difficulties in connecting to any of its databases. The MEDLARS Service Desk is available through a toll-free telephone number during weekday business hours as well as

---

[1]There are exceptions to the requirement for user IDs. Institutions and organizations contracting with NLM to provide networked access to the NLM databases can develop their own policies on user IDs and billing procedures for their users.

[2]Access to the NLM databases is available to the general public. Interested individuals can register for an NLM password and User ID code and can access the databases through a number of available mechanisms including direct modem connections, the Internet, or commercial or library database networks.

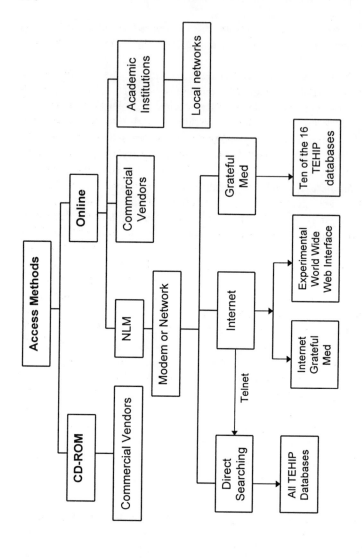

FIGURE 6.1 Primary access points to the TEHIP databases.

through email. In addition, the Regional Medical Libraries of the National Network of Libraries of Medicine provide technical support and assistance in searching. The committee commends NLM for these efforts and believes that increased publicity regarding the availability of this assistance may encourage more health professionals, especially those without networked access to attempt to search the NLM databases.

Cost

The NLM databases are available under a number of different pricing structures. NLM makes its databases available online by connecting directly to the NLM servers, which is one of the least expensive search options. In addition, NLM leases its databases to commercial businesses, which may repackage the databases in online or CD-ROM format. NLM also leases its database tapes to academic institutions and professional organizations, which incorporate the NLM databases into their internal computer networks and determine their own pricing structures. In an effort to expand its accessibility, NLM is offering organizations a new option: a fixed-fee access rate for Internet searching of the NLM databases (Wallingford et al., 1996).

A cost issue specifically affecting access to the TEHIP databases is the wide cost variations between the royalty and nonroyalty databases. TOXLIT, TOXLIT65, and CHEMLINE incorporate information from the Chemical Abstracts Service which charges royalties for usage. All other TEHIP databases and other MEDLARS databases are nonroyalty, and fees are assessed at the same rates for these databases.[3] Examples of cost variations include the citation charge (i.e., the cost per citation viewed, printed, or downloaded). The nonroyalty databases charge $.01 per citation, whereas the charges for TOXLIT and TOXLIT65 are $1.15 per citation, and the charges for CHEMLINE are $.69 per citation (NLM, 1996a).[4] The charges for each search statement submitted show similar variations: $.06 for nonroyalty databases, $1.78 for TOXLIT and TOXLIT65, and $1.07 for CHEMLINE (NLM, 1996a). It is important that cost issues be considered in examining the barriers to use of the TEHIP databases.

NAVIGATING THE TEHIP DATABASES

The complexities involved in navigating a database system are not unique to NLM. There is no standardization of search and retrieval methods, and each

---

[3]Note that there are no charges for the use of NLM's AIDS-related databases (see Chapter 2).
[4]Online charges listed are for regular billed domestic (U.S.) access.

> **BOX 6.2**
> **TOXNET Selection Menu**
>
> National Library of Medicine's Database Selection Menu
> Which TOXNET database would you like to access:
>
> [1] Hazardous Substances Data Bank - [HSDB]
> [2] Registry of Toxic Effects of Chemical Substances - [RTECS]
> [3] Chemical Carcinogenesis Research Info Sys - [CCRIS]
> [4] Integrated Risk Information System - [IRIS]
> [5] GENE-TOX - [GENETOX]
> [6] Environmental Mutagen Information Center Backfile - [EMICBACK]
> [7] Environmental Mutagen Information Center - [EMIC]
> [8] Environmental Teratology Information Center Backfile - [ETICBACK]
> [9] Developmental and Reproductive Toxicology - [DART]
> [10] TRIFACTS - [TRIFACTS]
> [11] Toxic Chemical Release Inventory Files - [TRI]

database developer produces a search system with unique features and its own degree of user-friendliness.

Selecting a Database

TEHIP users may have difficulty accessing the desired information because the content of each database is not immediately apparent from the initial TOXNET screens. NLM's present configuration of the TEHIP databases displays a list of databases from which to choose when logging onto the TOXNET system via the direct search mode (Box 6.2) or Grateful Med. Yet, the 16 TEHIP databases vary widely in the scope of their subject content and in the level of detail of the information. The name of the database provides only limited clues to its relevance to the user's information query. Additionally, because some of the TEHIP databases (TOXLINE, TOXLIT, DIRLINE, ChemID, and CHEMLINE) are available through the ELHILL server they are not listed on the TOXNET menu, and users may be unaware of the gateway between the ELHILL and TOXNET servers.

Although the committee realizes that frequent users know which database they need to access, numerous other potential users, including many health professionals, reach the list of databases and do not know in which direction to proceed. Efforts to assist users in determining which database to search by providing information on the scope of the subject matter and the type of information available (e.g., summaries, experimental data, and bibliographic citations) would be beneficial. This could be accomplished by the development of a decision-tree structure that would walk users through selection criteria, thereby simplifying this process for novice as well as experienced users (this issue is further discussed in Chapter 7).

Scope of the Databases

As described in Chapter 2, a wide range of information that may be useful to health provessionals is available in the TEHIP databases. The program evolved from a number of separate legislative mandates, interagency cooperative agreements, and NLM initiatives without the benefit from the beginning of a long-range plan. As a result, the content of the information in the databases overlaps and is somewhat uneven.

The NLM Long Range Planning Panel on Toxicology and Environmental Health (NLM, 1993) noted areas for expanded subject coverage in the TEHIP program. These included morbidity and mortality statistics related to environmental health, community emergency preparedness and response, radiation, and molecular medicine (NLM, 1993). Similarly, the IOM committee solicited input from health professionals regarding directions in which to expand the subject coverage of the TEHIP databases. The results indicate the need for additional information on occupational exposures and risk characterization; requests for expanded coverage of international research, epidemiologic and demographic research, and health outcomes data; and the exploration of the possibility of integrating geographic information system tools as part of the data systems.

The IOM committee did not explore the economic and interagency implications of supplementing the TEHIP databases with additional information or complementary databases. However, with the refinement of new technologies including the Internet, NLM may not need to directly provide the databases on its server but could provide pointers and linkages to other relevant databases (see Chapter 7).

User Interface

The balancing act in designing a user interface is to make it intuitive and easy to understand while still offering the user the power and depth of all of the features of the database program. Computer users are increasingly demanding interfaces that include simplicity while performing complex tasks; designs that work equally well for novice searchers as well as experts; graphics and text; customization options; and the ability to use the keyboard, mouse, pens, touch screens, and voice recognition (Alberico and Micco, 1990). Focus group participants and respondents to the committee's questionnaire indicated that a primary barrier to searching the TEHIP databases was the user interface.

For many of the TEHIP databases the user has several options in choosing an interface. The following section notes the strengths and weaknesses of each of the NLM search interfaces.

Direct Searching

As described in Box 6.1, direct searching is the only fully operational method of searching all of the TEHIP databases. Direct searching uses a command-line interface in which the user enters a search statement and receives input on the results. Command-line searching was developed for expert searchers who used early online bibliographic systems (Hersh, 1996). Consequently, this interface presents only skeletal directions, and the user must know the commands, the correct syntax, and Boolean logic to develop a targeted and precise search strategy (Box 6.3). The major disadvantage for novice and infrequent users of direct searching is its complexity and the steep learning curve needed to learn and master the commands and logic. Furthermore, extensive training or documentation is required to use the commands effectively and to determine the precise ways of keying in the search strategy.

Menu Searching

In an effort to simplify searching and thereby assist infrequent searchers, a menu interface has been developed for two of the TEHIP databases, TRI and HSDB. Menu searching provides the user with step-by-step screens that prompt the user in developing a search strategy and then modifying the strategy as needed. The initial TRI menu gives the user six choices (Box 6.4) and then continues to provide menus to assist the user in selecting the parameters for the search.

For many health professionals and other user communities who are interested in searching the TEHIP databases but who do not have the time to learn the command language, menu searching offers a preferred search interface. The trade-off with the menu system is the loss of the intricacies of direct searching, including the ability to modify the search quickly and to narrow or expand the search strategy easily.

Grateful Med

The user interface for Grateful Med was designed by NLM for use by health professionals and other end users who do not have the time or inclination to learn the syntax and commands needed for direct searching (Hersh, 1996).

BOX 6.3
Sample Search on HSDB

[HSDB] SS1/cf?	Indicates that the user should enter the first search statement
User: **(USE) solvent**	First search statement requesting records for chemicals used as solvents
Search in progress SS(1) PSTG (468)	System message indicating 468 records match the first request
[HSDB] SS2/cf? User: **(htox) spinal and cord**	Second search statement requesting records with both the words "spinal" and "cord" in the human toxicity field
Search in progress SS(2)PSTG (32)	System message indicating 32 records match the second search statement
[HSDB] SS3/cf? USER: **1 and 2**	Requests that records that meet the search requirements of both searches 1 and 2 be retrieved
Search in progress SS(3) PSTG (2)	Indicates that there are two records with information on solvents affecting the spinal cord
[HSDB] SS4/cf? USER: **prt htox**	Print command; indicates that the human toxicity (htox) field should be printed

NAME – Diethyl ether
RN – 60-29-7
HTOX – The principal physiological effect is anesthesia. Repeated exposures in excess of 400 ppm may cause nasal irritation, loss of appetite, headache, dizziness, and excitation, followed by sleepiness. Mental disorders have been reported, as has kidney damage. [Encyc Occupat Health & Safety 1983, p. 786] **Peer-reviewed**
HTOX – Repeated contact with the skin may cause it to become dry and cracked. [Encyc Occupat Health & Safety 1983, p. 786] ** Peer-reviewed**
HTOX – Cases of human death in industry due to acute inhalation are rare. One such case, subject developed acute mania and died in uremic convulsions. [Patty. Indus Hyg & Tox, 3rd ed. Vol. 2A, 2B, 2C. 1981-82. p. 2508] ** Peer-reviewed**
HTOX – Clinically, albumin may appear in the urine, and polycythemia in blood. Nephritis may develop in rare cases. [Patty. Indus Hyg & Tox, 3rd ed. Vol. 2A, 2B, 2C. 1981-82. p. 2509] ** Peer-reviewed**

NOTE: This is only a partial printout of the search results. **Boldface type** indicates input by the searcher.

> **BOX 6.4**
> **Initial TRI Menu**
>
> Do you wish to search the Toxic Chemical Release Inventory by:
>
> [A] Facility Location?
> [B] Facility Identification?
> [C] Chemical Identification?
> [D] Publicly Owned Treatment Works Identification?
> [E] Other Off-Site Locations Identification?
> [F] Environmental Releases?

This program simplifies the search process for novice users while also offering the option of direct searching for experienced searchers. The user can input terms or may access the MeSH thesaurus and select the appropriate MeSH terms. The Grateful Med program automatically inserts the Boolean logical operators into the search request and handles other technicalities (e.g., formatting the author's name correctly. As with menu searching, expert searchers may find Grateful Med limiting, whereas novice searchers get useful results with minimal effort or training.

Internet Grateful Med debuted in April 1996 and is available for searching MEDLINE, HealthSTAR, AIDSLINE, and PREMEDLINE. The interface is user friendly (buttons and pull-down selection boxes are used) and offers a number of ways to limit a search (e.g., by type of publication, date, or language). One valuable addition to this interface is the incorporation of the Unified Medical Language System Metathesaurus (discussed below), which analyzes the search strategy and suggests additional terms that could be used to expand or narrow the search strategy.

Experimental World Wide Web Search Interface

An experimental approach to searching the TEHIP databases is available to users on the World Wide Web (WWW) through the Special Information Services (SIS) homepage (http://sis.nlm.nih.gov). There are two search interfaces in prototype form: Webline, which is designed to search the bibliographic files MEDLINE, TOXLINE, and AIDSLINE; and the TOXNET Experimental WWW Search Interface, which searches the TEHIP factual databases. Both interfaces use hypertext linkages, a point-and-click interface, and pull-down selection boxes.

The initial screen for the TOXNET Experimental WWW Search Interface provides choices of subject areas that can be searched: health effects, emergency medical treatment, or scientific literature (Box 6.5). A search on all of the fac-

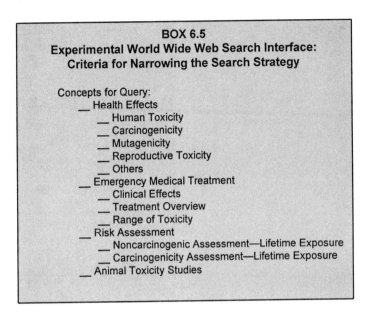

BOX 6.5
Experimental World Wide Web Search Interface:
Criteria for Narrowing the Search Strategy

Concepts for Query:
__ Health Effects
 __ Human Toxicity
 __ Carcinogenicity
 __ Mutagenicity
 __ Reproductive Toxicity
 __ Others
__ Emergency Medical Treatment
 __ Clinical Effects
 __ Treatment Overview
 __ Range of Toxicity
__ Risk Assessment
 __ Noncarcinogenic Assessment—Lifetime Exposure
 __ Carcinogenicity Assessment—Lifetime Exposure
__ Animal Toxicity Studies

tual TEHIP databases is conducted, and it is only when the search results are presented that the searcher becomes aware of the multiple databases that were searched. This interface presents choices that are of interest to health professionals (e.g., emergency medical treatment) and does not require special training or extensive documentation.

The disadvantages of this search method are not a result of the user interface but, rather, are a question of access to the Internet. Although a recent survey of 2,500 searchers of the NLM databases found that 75 percent have Internet access, 36 percent of rural users and almost half of users at hospitals indicated that they do not yet have access to the Internet (NLM, 1996b). Because access to the Internet is growing rapidly, this may not be a concern in the near future. However, it will be important to coordinate the development of this interface with Internet Grateful Med to produce a single interface (or similar interfaces) for searching all of the NLM databases on the Internet.

Conducting a Search

Searchers begin a database search with a specific query or information need. Focus group participants stated that many of their queries for toxicology and environmental health information begin with a specific chemical name. However, they also discussed the need for other starting points for their queries that fall into three general categories:

- exposure-related queries, particularly those based on a job description or a geographic location,
- symptom-related queries, and
- regulation-related queries.

Additionally, primary care professionals, emergency medical personnel, and poison control center personnel need to be able to conduct searches on the basis of the brand names of household or other products to retrieve treatment information. Commercially-available databases, such as POISINDEX®, provide brand name searching and interfaces that are developed to provide rapid access to the acute care information needed by poison control centers and emergency departments.

One of the hallmarks of online databases has been the use of controlled vocabulary indexing to provide users with a tool to increase the precision of their search strategies (Siegel et al., 1990). NLM has been a leader in this field with the development and continual refinement of the MeSH thesaurus. Many health professionals are familiar with MeSH through searching the MEDLINE database and may assume that all of the NLM databases are indexed with MeSH terms. However, that is not the case, due to the diversity of the originating federal agencies responsible for the TEHIP databases.

The TEHIP databases are designed to facilitate chemical-related searches, but they are less well-equipped to handle searches from other entry points. The field that is common to all of the TEHIP factual databases except DIRLINE is the CAS Registry Number field. NLM has recently added the capacity to perform chemical-based cross-file searches across six of the databases (HSDB, CCRIS, GENE-TOX, IRIS, RTECS, and TRIFACTS) by linking the files through the Registry Number field. Additionally, free-text searching allows the searcher access to words in the abstract and other text fields; however, indexing across the databases is not consistent. Although this issue may present a barrier for health professionals using the TEHIP databases, a simple cost-effective solution is not available. Indexing is a labor-intensive process, and the cost of adding MeSH terms to the TEHIP databases would be prohibitive (Siegel et al., 1990). The new technologies discussed at the end of this chapter, including natural language searching, should ease current constraints and should, in the near future, facilitate effective searching from a variety of entry points.

Evaluating and Refining the Search: Quality Indicators

The last step in the search process is an evaluation of the search results and a refinement of the search strategy, if needed. One of the factors involved is an evaluation of the validity and reliability (i.e., quality) of the data. Health professionals, through input provided by focus groups and questionnaires, expressed the need for clear indicators of the level of scientific review on all data retrieved from the TEHIP factual databases. This is particularly important for summarized information when the searcher is not presented with the original data. Focus groups brought up the additional concern that there are widely differing scientific opinions on the toxicities of certain chemicals, in part because of the paucity of data from studies with humans and difficulties in extrapolations from animal studies. These different opinions point to the need for peer review and for clear indications of the source and quality of the data.

The issue of quality indicators was discussed in the NLM Long Range Planning Panel report (NLM, 1993), and NLM has made efforts to address this concern. As described in Chapter 2 and summarized in Table 6.1, the levels of review differ between databases. To indicate the extent of review, NLM has added statements that are displayed after the user has logged onto the NLM system and has selected a database. For example, after selecting the GENE-TOX database, the user receives the message, "The mutagenicity data in this file have been peer-reviewed." Each data element in HSDB is marked with one of three tags indicating the level of review: "No review," "QC reviewed," meaning that the data have been reviewed for quality errors such as missing or contradictory data, or "Peer-reviewed," meaning that HSDB's Scientific Review Panel has reviewed this information. In many of the databases the data are not reviewed for content but have been internally validated to confirm that there were no input errors.

Indicators of the extent of peer review and other quality control measures are of utmost importance to the medical and scientific communities. It would be advantageous for searchers to have as an option the ability to limit their searches to only peer-reviewed information. This option could be considered a component of new interfaces such as those being developed on the World Wide Web.

TABLE 6.1 Review Process for TEHIP Factual Databases

Peer-Reviewed by Reviewers External to the Originating Agency	Scientifically Reviewed by Reviewers in the Originating Agency	Data Validation Checks
HSDB	CCRIS[a]	ChemID
GENE-TOX	IRIS	CHEMLINE
		DIRLINE
		RTECS
		TRI
		TRIFACTS

[a]Experts in carcinogenicity and mutagenicity select sources and provide scientific evaluation.

FUTURE DIRECTIONS

Effective Interfaces

Effective user interface design incorporates the psychological, behavioral, and technological factors that are involved in the human-machine interaction (human factors engineering) (Kohoutek, 1992). The committee received input from focus group participants, questionnaire respondents, invited speakers, and discussions with colleagues on the issues involved in simplifying the current user interface to the TEHIP databases with the goal of achieving an interface that assumes no a priori knowledge on the part of the user with respect to the data sources, their format, or content. The committee realizes that the state-of-the-science in interface design is rapidly changing and chooses to highlight a few of the innovations that could enhance the TEHIP interface:

- *transparent or seamless interface between databases* (an interface in which the user is unaware of the complexity of the database system underlying the search);
- *graphics-based interface* (an interface that uses buttons, icons, pull-down selection boxes, and other methods of inputting searches [e.g., Internet Grateful Med and the TEHIP Experimental World Wide Web interface]);
- *relevance searching* (the information retrieved is ranked in order of relevance to the search strategy);
- *query by example* (the searcher selects one reference that is on target, and the system then searches for all references similar to the example); and
- *search analysis* (the search program analyzes the search strategy and suggests terms that could narrow or expand the search [e.g., Internet Grateful Med]).

Access

As previously discussed, users can access the entire TEHIP complement of 16 databases in only one way: direct searching. However, direct searching is the most difficult interface for novice users due to its complex command-line interface. SIS is developing a more intuitive search interface using the graphical and hypertext capabilities of the Internet World Wide Web. The committee believes that this effort would provide an attractive search option for health professionals. However, the committee hopes that this effort will be coordinated with the development of Internet Grateful Med and other future Internet developments at NLM, to provide health professionals with similar interfaces whether they are searching MEDLINE or one of the TEHIP databases.

The rapid expansion of the World Wide Web has also created possibilities for links from other sites to the TEHIP Web site and has led to the development of new information dissemination approaches for health professionals. One example is Physicians' Online, a Web-based resource sponsored by pharmaceutical companies that provides free access to several NLM databases (MEDLINE, AIDSLINE, CANCERLIT, and HEALTH) for all physicians and other health care professionals who write prescriptions. Adding free access to the more clinically-relevant TEHIP databases (e.g., HSDB and TOXLINE) would promote use by a number of health professionals who are not familiar with these databases. Although these decisions are not in NLM's purview, NLM could explore these as possible access points. Additionally, NLM could explore the possibilities of providing hypertext linkages between the World Wide Web homepages of health professional organizations and the TEHIP Web site and the possibility of using Web search engines to provide pointers to TEHIP's Web site.

Search Vocabularies

One of the stumbling points for database users has been the fact that many databases or database systems use their own unique thesaurus of keywords or descriptors. If the user wants to perform a comprehensive search, he or she must become familiar with the search vocabulary, or if the database does not use a controlled vocabulary, the searcher must determine all relevant synonyms and related terms on which to search. To assist users in overcoming these quandaries, NLM has initiated an extensive research and development project, the Unified Medical Language System (UMLS). The basic assumptions of the UMLS project are that databases will continue to have diverse vocabularies and that users will continue to use a variety of terms in searching for biomedical information (Humphreys and Lindberg, 1993). The UMLS project has developed four knowledge sources, three of which (the Metathesaurus, Semantic Network, and SPECIALIST lexicon) integrate diverse vocabularies by linking terms on the basis of conceptual, semantic, and lexical connections, and the fourth, the Information Sources Map, is used to link the user with the appropriate information resource (McCray et al., 1993, 1994; Schuyler and Hole, 1993; see also Chapter 7). The UMLS knowledge sources are available to systems developers, who can incorporate them into expert systems that provide cross-database searching, natural language queries, and a variety of other applications.

The UMLS knowledge sources have been incorporated into the Internet Grateful Med program and provide users with additional terms that can be used to narrow or expand the search strategy. The UMLS project is especially applicable to the TEHIP databases because these databases are not indexed with

MeSH terms or any other single controlled vocabulary. Consequently, efforts to facilitate free text and natural language searching will result in increased search effectiveness. This will be particularly useful for nonchemical-based search strategies such as searching by symptoms or queries on exposure regulations. Applying this technology to the TEHIP databases will assist users in constructing effective search strategies and will result in increasingly precise search retrieval results.

CONCLUSIONS AND RECOMMENDATIONS

The issues involved in accessing and navigating the TEHIP databases are in a state of flux because of rapidly changing computer technology and the expansion of the Internet. However, the committee wishes to stress that the TEHIP databases should be included in the technological changes of the future. The pioneering efforts applied to MEDLINE should, when applicable, be incorporated into the TEHIP program. As discussed in Chapter 1, the demands for toxicology and environmental health information by health professionals and other interested user communities, including the general public, are increasing, and it is critically important for information resources to be easily accessible with intuitive search interfaces.

The committee recommends that a two-step approach be implemented in making refinements to the TEHIP databases and emphasizes the need for evaluation components to be incorporated and then monitored throughout this process. The first step would entail improvements that could be made in the short term while the user analysis (recommended in Chapter 4) is being conducted. The second step would be implemented over the long term and would be based on the results of the user analysis. Once the results of the user analysis are examined and it has been determined which of the TEHIP databases are most useful to health professionals, then a prioritization of activities (including outreach and training, access, and navigation) should be undertaken around those most useful databases. The committee's short- and long-term recommendations with regard to access and navigation are detailed below.

The committee recommends that in the short term (during the time that the user analysis is being conducted) NLM continue its efforts to increase access to the TEHIP databases and to simplify navigation of the databases by:

- **coordinating the development of the TEHIP Experimental World Wide Web Interface with Internet Grateful Med,**
- **promoting online registration for database access, and**

- exploring the possibilities of linking TEHIP's World Wide Web site with the Web sites of other health professional organizations and establishing pointers to the TEHIP databases from World Wide Web search engines.

The committee recommends that in the long term and on the basis of priorities set as a result of the user analysis, NLM expand its efforts to facilitate access and navigation of the TEHIP databases by making full use of the navigational tools being developed within NLM and beyond. This includes:

- implementation of more efficient and intuitive user interfaces,
- incorporation of UMLS knowledge sources and other expert systems that would enhance symptom-related and other natural language searches,
- incorporation of navigational tools and interfaces that would create a seamless interface between databases, and
- implementation of indicators of peer review into new search interfaces.

REFERENCES

Alberico R, Micco M. 1990. *Expert Systems for Reference and Information Retrieval.* Westport, CT: Meckler.

Hersh WR. 1996. *Information Retrieval: A Health Care Perspective.* New York: Springer-Verlag.

Humphreys BL, Lindberg DAB. 1993. The UMLS project: Making the conceptual connection between users and the information they need. *Bulletin of the Medical Library Association* 81(2):170–177.

Kohoutek HJ. 1992. Assuring quality of the human-computer interface. *Quality and Reliability Engineering International* 8(5):427–440.

McCray AT, Aronson AR, Browne AC, Rindflesch TC. 1993. UMLS knowledge for biomedical language processing. *Bulletin of the Medical Library Association* 81(2):184–194.

McCray AT, Srinivasan S, Browne AC. 1994. Lexical methods for managing variations in biomedical terminologies. *Proceedings of the 18th Annual Symposium on Computer Applications in Medical Care.* Pp. 235–239.

NLM (National Library of Medicine). 1993. *Improving Toxicology and Environmental Health Information Services.* Report of the Board of Regents Long Range Planning Panel on Toxicology and Environmental Health. NIH Publication No. 94-3486. Bethesda, MD: NLM.

NLM. 1996a. *NLM Online Charges* [http://www.nlm.nih.gov/databases/us_price.html]. December.

NLM. 1996b. *Survey of Online Customers: Usage Patterns and Internet Readiness.* NIH Publication No. 96-4181. Bethesda, MD: NLM.

Schuyler PL, Hole WT. 1993. The UMLS Metathesaurus: Representing different views of biomedical concepts. *Bulletin of the Medical Library Association* 81(2):217–222.

Siegel ER, Cummings MM, Woodsmall RM. 1990. Bibliographic retrieval systems. In: Shortliffe EH, Perreault LE, eds. *Medical Informatics: Computer Applications in Health Care.* Reading, MA: Addison-Wesley.

Wallingford KT, Ruffin AB, Ginter KA, Spann ML, Johnson FE, Dutcher GA, Mehnert R, Nash DL, Bridgers JW, Lyon BJ, Siegel ER, Roderer NK. 1996. Outreach activities of the National Library of Medicine: A five-year review. *Bulletin of the Medical Library Association* 84(2 Suppl).

7

Program Issues and Future Directions

In this concluding chapter, the committee highlights four areas that are important to the future growth and development of the TEHIP program: stable funding, the integration of the TEHIP program into NLM initiatives, the establishment of an advisory committee, and the implementation of a strong evaluation process. Additionally, the committee presents its recommendations for future directions that will provide health professionals with the tools necessary to readily access toxicology and environmental health information. The committee reiterates its belief that NLM should continue to exert its role as a library and, in doing so, expand awareness of the full range of national, state, local, international, and private-sector information resources in toxicology and environmental health (see Chapter 3). The continued challenge for the TEHIP program will be to provide easily accessible and authoritative information to an increasingly computer literate and environmentally-aware user community that includes health professionals and the general public.

The 1993 report by the NLM Long Range Planning Panel on Toxicology and Environmental Health (NLM, 1993) served as an invaluable resource to the committee. The committee found the report to be a useful basis for its work and wishes to reaffirm the planning panel's recommendations, some of which have not yet been implemented. It is the committee's hope that NLM will reexamine the planning panel's report and use its recommendations in conjunction with the recommendations of this report to strengthen the TEHIP program.

PROGRAM ISSUES

Funding

A critical factor to the success of any program is a stable level of funding adequate enough to maintain staffing levels and meet programmatic requirements. As with NLM as a whole, the TEHIP program receives funds from two sources: funds legislatively appropriated to NLM and reimbursements from other government agencies.

As discussed in Chapter 2, the TEHIP program's budgetary appropriations have remained fairly constant over the past 29 years; however, fluctuations in reimbursements from other agencies have been significant. Reimbursement funding is the result of collaborative projects with other federal agencies, and the TEHIP program is funded in large part through interagency agreements. However, changing priorities, responsibilities, and resources within the various agencies have significantly affected the TEHIP program. Between 1992 and 1993, the reimbursable budget dropped by approximately 50 percent, from $2.45 million in FY 1992 to $1.27 million in FY 1993. Since 1993, the reimbursable budget has remained at that reduced level (the FY 1995 reimbursable budget was $1.23 million).

For the TEHIP program to be responsive to the changing demands of health professionals and to improve the utility of the databases, funding must be adequate to implement changes based on the results of the user profile analysis (Chapter 4) and on committee recommendations for improving access, navigation, and program evaluation. Fluctuating funding makes it difficult to plan future activities, to develop long-range goals, and to implement necessary program changes. As noted, however, a mechanism for using limited resources is the prioritization effort based on the results of the user analysis. Nonetheless, fluctuating funding will affect staffing and the ability to keep pace with the rapidly changing trends in technology necessary to make the most useful databases even more accessible to health professionals. The committee did not discuss what appropriate levels of funding might be and suggests that NLM determine the level of commitment to the TEHIP program and develop a cost analysis based on the needed improvements to the program. Clearly, the TEHIP program requires a stable funding base that is not subject to changes in the priorities and programs of other federal agencies.

Leadership Role for TEHIP

The committee has noted the need for a strong leadership role in promoting and marketing the TEHIP program. As NLM engages in new initiatives and research and development (R&D) projects that are relevant to the TEHIP program, especially with regard to state-of-the-art computer technology, the TEHIP program leadership should urge NLM to integrate the TEHIP program into those initiatives. For example, as NLM continues to develop the Unified Medical Language System, Internet Grateful Med, and a range of training and outreach programs, TEHIP's priorities should also be included in those activities. Thus, the TEHIP program leadership should be vocal about its program priorities within NLM so that they may be incorporated into new initiatives and be addressed by NLM. Similarly, the committee encourages NLM to more fully integrate the TEHIP program's activities into broader NLM efforts in order to maximize their effects, avoid duplicative efforts, and leverage limited resources.

The mission statement of the TEHIP program has three priorities: to provide selected core information resources and services in toxicology and environmental health, to facilitate access to national and international information resources in this field, and to strengthen the information infrastructure of toxicology and environmental health. The committee believes that the TEHIP program is well-positioned to carry out its mission, which incorporates the committee's recommendation for broadening the focus of the program to encompass the full range of toxicology and environmental health information. The complexity of scientific and technical issues faced in implementing the TEHIP program requires a knowledgeable staff with a range of scientific and computer expertise. The longevity of the TEHIP program is due in large part to the high level of expertise and dedication of the Specialized Information Services Division (SIS) staff. The committee believes that the TEHIP program has demonstrated a commitment to develop, maintain, and improve environmental health databases. As TEHIP staff continue to refine and improve the databases and fulfill the mission, as interpreted broadly by the committee, they should exert their leadership role in providing environmental health and toxicology information by incorporating their program priorities into NLM activities where appropriate.

Advisory Committee

The growth and development of any program are strengthened and invigorated by advice provided by outside experts. NLM uses this mechanism through a number of boards and advisory committees. Four of the six divisions within NLM's organizational framework have standing advisory committees made up of external scientists, librarians, and health professionals (Figure 7.1). In some cases the committees have specific tasks; for example, the Literature Selection Technical Review Committee selects journals for indexing in *Index Medicus* and MEDLINE, and the Biomedical Library Review Committee supports the work

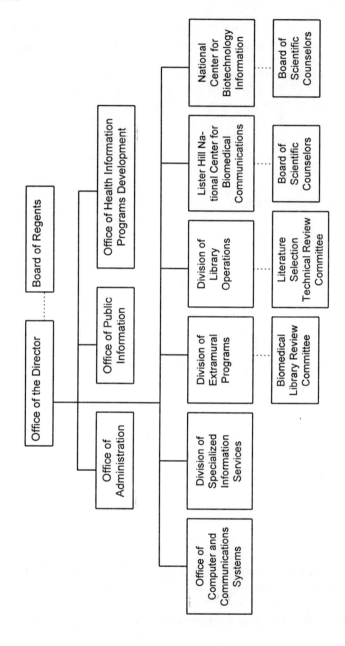

FIGURE 7.1 NLM advisory committees.

of the Division of Extramural Programs by reviewing grant applications under the Medical Library Assistance Act. The Lister Hill National Center for Biomedical Communications and the National Center for Biotechnology Information each have a Board of Scientific Counselors that reviews and makes recommendations on NLM's intramural research and development program (NLM, 1995).

For 25 years (1969 to 1994), SIS received policy and technical advice from an advisory committee of the National Academy of Sciences (NAS), the Toxicology Information Program Committee (TIPCOM). The committee was made up of leading toxicologists, pharmacologists, chemists, and computer scientists who reviewed the progress and growth of the toxicology and environmental health program at NLM. In 1993, NLM's focus turned toward examining the medical relevance of the TEHIP databases and their under-utilization by health professionals. NLM requested that NAS, through the Institute of Medicine (IOM), convene a planning meeting to examine health professionals' use of the TEHIP databases and suggest future evaluation activities. Subsequent to the planning meeting, NLM and IOM developed the study on which this report is based. There is currently no other advisory committee for the TEHIP program or for the other diverse activities of SIS in general.

The committee has considered the needs for continuing oversight of the TEHIP program and believes that an outside advisory committee would benefit the program. SIS has multiple responsibilities that are primarily split between two subject areas: toxicology and environmental health and HIV/AIDS. These two areas differ widely in the issues that need to be addressed and in the interested communities and advocacy groups. Therefore, the committee believes that an advisory committee should be established specifically for the TEHIP program.

A necessary component of any advisory committee is stakeholder participation. A large number of diverse groups are interested in the TEHIP program, and their participation and input would be beneficial. These groups include toxicologists (medical and clinical), health science librarians, medical informaticians, computer scientists, environmental scientists, environmental policymakers, health care professionals (including tertiary care professionals in occupational and environmental health; primary care professionals; and emergency medicine, public health, and poison control personnel), and interested members of the general public including community environmental activist organizations and public librarians.

Additionally, the committee suggests that SIS obtain input via other mechanisms on a periodic basis. This may include the use of focus groups, the formation of a liaison panel to the advisory committee composed of representatives from other federal agencies involved in environmental health issues, or the use of other evaluation mechanisms such as surveys and workshops.

To cover all relevant issues, the advisory committee should examine the issues brought to its attention by NLM, by advisory committee members, and by other interested parties. It is important that the advisory committee provide timely and relevant advice on issues, including the following:

- program content priorities, management, planning, and evaluation;[1]
- resource needs;
- quality assurance and peer review;
- relationships with other government agencies and initiatives; and
- technical operations.

The outside advisory body should be constituted to facilitate a two-way flow of information and ideas. Beyond its role of providing advice to the program, there is a significant role for the advisory committee in conveying knowledge about the program and its products back to the range of communities represented on the committee. This may include presentations at professional conferences and involvement in outreach activities.

Evaluation

An integral component to the effectiveness of any program is a thorough and ongoing evaluation effort planned during the initial phase of each project and incorporating benchmarks to assess ongoing progress. NLM has used a range of methods and strategies to assess the effectiveness of its products, services, programs, and policies; however there is a need for expanded evaluation efforts.

Current evaluation methods include beta-testing, a process of iterative testing and refinement that uses input provided from groups of end users. Selected test sites, often NN/LM member libraries, implement the new product and collect user input. For example, the beta-test version of the LOANSOME DOC[2] program, software, and documentation was sent to 57 libraries and more than 600 health professionals (Wallingford et al., 1996). Additionally, user input has been sought through surveys, workshops, and focus groups. Needs assessment and baseline surveys have been used to assess the level of knowledge about NLM's products (e.g., Grateful Med) prior to an outreach effort, the impact of which is then evaluated by follow-up questionnaires. One of the new mechanisms in evaluation efforts is the use of focus groups in which small numbers of people, often 10 to 12 peers, discuss specific topics through the leadership of a

---

[1] Overall evaluation of the TEHIP program should include an examination of priorities, the incorporation of technological improvements, and the merging of resources when and where possible.

[2] LOANSOME DOC is a software program used to order the full text of articles retrieved from references in MEDLINE and other NLM databases.

facilitator (Seiver, 1989). Focus groups of health professionals have been used to assess familiarity with NLM services and barriers to their use (Mullaly-Quijas et al., 1994).

Consideration should be given to developing and implementing an evaluation plan focused specifically on the TEHIP program. This may include the detailed user analysis (recommended in Chapter 4) and the previously mentioned advisory committee. Evaluation is often considered only when the program or product is in place or fully developed. Thus, the committee emphasizes the importance of incorporating an evaluation component into a project during its initial planning stages and early implementation. Additionally, the committee stresses the advantage of building benchmarks into the project plan to assess progress at specified stages. Evaluation studies for new TEHIP products and for training and outreach efforts should continue to use a variety of traditional approaches (including surveys and focus groups) and should also expand to use new methods (e.g., Internet access counters, online surveys). Input should be sought from health professionals in a range of work settings and with a range of expertise.

Recommendation on Program Issues

As the committee conducted this study, several programmatic issues came to the forefront. It is clear that NLM has taken a great deal of initiative in disseminating toxicology and environmental health information; however, there is a need for a stabilized funding base, an internal commitment to the TEHIP program, an involvement of the TEHIP program in broader NLM R&D efforts, an interdisciplinary advisory committee, and the development and implementation of an evaluation plan. The TEHIP program continues to make a substantial contribution to the fields of toxicology and environmental health, and the beneficial use of toxicology and environmental health information resources by health professionals and other interested user communities can be considerably increased given the necessary resources and support.

The committee recommends that the TEHIP program be given the responsibilities and resources needed to strengthen its growth and development. This may be accomplished by:

- providing a stable funding base,
- ensuring a leadership role for the TEHIP program and promoting the incorporation of the TEHIP program into broader technological developments at NLM,
- establishing an interdisciplinary advisory committee, and
- implementation of an evaluation plan.

FUTURE DIRECTIONS

Because health professionals will continue to need accessible and authoritative sources of toxicology and environmental health information, the committee believes that it is important to provide health professionals with the tools to retrieve this information. During the course of its deliberations, the committee determined that the optimal approach would be to provide health professionals with two different options for accessing toxicology and environmental health information. Primary access would be via an online directory or "road map" that would assist the user in identifying and connecting with the relevant TEHIP database or other (non-TEHIP) online information resource (e.g., World Wide Web sites of government agencies, directory information for environmental health organizations).[3] Ultimately, the search would be conducted automatically and seamlessly without requiring the user to select specific information resources. The second access option would be a toll-free telephone number or similar single-access information center that would link health professionals (person-to-person) with a specialist who could provide consultative services on environmental health issues and concerns.

This two-part approach would require cooperative efforts between NLM, other federal agencies, and private-sector organizations. NLM's expertise and current R&D efforts in medical informatics would make it the logical agency to take the leadership role for the online directory, although the assistance of other agencies and organizations would be needed to provide directory information. The development and implementation of the second option, the single-access information center, go beyond NLM's mandate, and this recommendation should be considered by multiple federal agencies and other private-sector organizations involved in environmental health. Public-private partnerships could play a key role in providing these information services. The collaboration of NLM with universities, industry, international resources, and local, state, and federal governments could be particularly productive given the numerous information resources in this field (Chapter 3).

Online Directory

The committee believes that an online directory of toxicology and environmental health information resources would be most useful if it not only provided information about the available online resources in toxicology and environmental health but also assisted the user in locating and connecting to the most

---

[3]The committee's vision of the online directory includes simple interactive screens that walk the user through various options for reaching the appropriate information resource through hypertext links, graphical interfaces, and search engines.

relevant resource for his or her information query. Currently, it can be time consuming to search the World Wide Web for information resources on environmental health, as users may experience the frustration of massive and poorly differentiated retrievals. As recommended in Chapter 3, it is important for NLM to exert a traditional library role in comprehensively assembling, organizing, and maintaining an inventory of information resources in toxicology and environmental health. The online directory would involve an extensive compilation with links to online databases; directories of environmental health specialists; Web sites of government agencies and private-sector organizations involved in environmental health; and other similar information resources.

The TEHIP program staff has started this process by assembling an Internet Web site that links the user to other federal and international Web sites with toxicology and environmental health information. Additionally, much of the directory information is available in the current DIRLINE database. To maximize the utility of this approach, an inventory of all relevant information resources in this field should be conducted and incorporated into the data management scheme.

An important component of this road map would be the incorporation of tools such as overlying decision trees that would assist the user in locating the relevant resource. Decision points could include questions regarding the purpose for which the information is needed (e.g., patient care, research, policy development, or education), the type of information needed (e.g., bibliographic, factual, or directory), the subject matter, or the type of information needed (e.g., animal data, epidemiologic studies, or clinical consultation). Additionally, for those users requiring expert assistance, the decision tree could provide pointers to the single-access information center (described in the next section). In order to build effective decision trees, it would be important for the online directory to incorporate information beyond pointers such as subject coverage, access points, and quality indicators. Input on the key elements to be included in the decision tree structure could be provided by the TEHIP program advisory committee and by the user profile analysis.

NLM is currently funding R&D efforts on the components of expert systems that could provide this decision tree and directory technology. The Unified Medical Language System (UMLS)—one of NLM's major research initiatives—is a long-term R&D effort that aims to assist health professionals and other searchers in retrieving and integrating online biomedical information from diverse information sources. Currently, the UMLS Information Sources Map (ISM) is a promising component of a system that would analyze a searcher's information request to identify and connect the user to the appropriate information source (Clyman et al., 1993). The ISM is a database with records that describe information providers and information sources. The scope of each information source is indexed by using MeSH and other relevant terms. NLM, in a collaborative project with the Clearinghouse for Networked Information Discovery and Retrieval, has developed a Web-based application, Sorcerer, that accepts user queries, identifies the information sources in the ISM that are likely to have relevant information, and then connects to the source, conducts the query, and returns the results to the user (NLM, 1995). A companion software tool, Apprentice, is in development. Apprentice will enable information providers throughout the biomedical community to register their information sources with the centralized ISM database (NLM, 1995). Expert systems in development and prototype phases are incorporating the ISM. This technology, in combination with the hypertext capabilities of the World Wide Web and other advances in informatics, bodes well for the online directory to be an attainable goal in the near future.

A model for the development of this online directory is the National Environmental Data Index (NEDI), an interagency effort to provide a central Internet directory or "yellow pages" for accessing environmental data (e.g., data on air and water quality, global change, or renewable energy; see also Chapter 3). Phase I of this project is in progress and aims to provide distributed access to data systems within the federal government. Phase 2 will expand to provide links to state, local, international, and private (nonprofit and commercial) data systems (NEDI, 1996). Links between NEDI and the proposed online directory for environmental health information would be critical for providing comprehensive information to all interested user communities.

Single-Access Information Center

In addition to a directory of online information resources, the committee became aware that in some clinical and other situations, health professionals would prefer to consult with an expert in environmental health. Although there are several toll-free telephone numbers that provide limited information related to environmental health and some poison control centers provide environmental health services, there is no single resource that health professionals or others can call with detailed questions on these topics. Many focus group participants expressed the need for a single-access point for toxicology and environmental health information, particularly clinically-pertinent information in real time.

This idea has been discussed and recommended by previous IOM committees that examined the occupational and environmental health information needs of primary care professionals. Those committees determined that a comprehensive center providing clinical and nonclinical services through a consultation structure would meet the information needs of many health professionals (IOM, 1988, 1990). The center could provide expertise in patient care, risk assessment, and exposure reduction, including industrial hygiene, diagnostic and treatment services, and assistance in dealing with government agencies (IOM, 1988). Justifications for a single-access point include the need for rapid and authoritative information, the scarcity of specialists in occupational and environmental health, the fragmentation of environmental health responsibilities throughout the agencies of the federal government, and the vast and disparate information resources in this area.

Poison control centers are examples of single-access points that are effectively meeting the information needs of health professionals and other user communities. Poison control centers provide 24-hour-a-day, 7-day-a-week service to health professionals and the general public primarily by responding to calls for immediate treatment management of toxic exposures. Additionally, poison control centers provide telephone follow-up and, as necessary, provide referrals for specialized health care. Forty-seven regional poison control centers are certified by the American Association of Poison Control Centers (AAPCC), and several of these have developed occupational and environmental health information services. AAPCC compiles the Toxic Exposure Surveillance System, a nationwide database on human exposure cases. Poison control centers fill a unique niche in emergency medical services, and it is estimated that at least $5 is saved in health care costs for each dollar spent on poison control center services (Durbin and Henretig, 1995; Kearney et al., 1995; Lovejoy et al., 1994). The savings are attributed primarily to decreases in the unnecessary use of ambulance and emergency department services for minimal symptoms. However, many centers are facing substantial budgetary cutbacks.

The implementation of a single-access information center would have budgetary and interagency ramifications that the committee did not have the mandate to explore. However, in its exploration of the TEHIP databases and its inquiries into the dissemination of toxicology and environmental health information—especially to health professionals—the committee realized the need for this type of information resource for toxicology and environmental health information. The committee believes that poison control centers, which are already established, would make excellent resources for providing toxicology and environmental health information to health professionals. However, the committee is particularly mindful of the budgetary considerations and of the problems that poison control centers would face if their mandate is expanded without the necessary financial resources for implementation. Although the provision of a single-access information center for toxicology and environmental health information is not within the purview of NLM, the issue should be explored because it is important to expanding the use of this information by health professionals.

The committee recommends that NLM, other relevant federal agencies, and private-sector organizations work cooperatively to provide health professionals and other interested user communities with the tools that they need to access toxicology and environmental health information. This would involve two different types of access points:

- an online directory that would contain information on the full spectrum of information resources in toxicology and environmental health and that would direct the user to the appropriate online information resource, and
- a single-access information center (e.g., regional poison control centers) that would connect the user with individuals with expertise in environmental health.

REFERENCES

Clyman JI, Powsner SM, Paton JA, Miller PL. 1993. Using a network menu and the UMLS Information Sources Map to facilitate access to online reference materials. *Bulletin of the Medical Library Association* 81(2):207–216.

Durbin D, Henretig F. 1995. Poison control center impact on emergency department utilization rate (abstract). *Clinical Toxicology* 33:557.

IOM (Institute of Medicine). 1988. *Role of the Primary Care Physician in Occupational and Environmental Medicine*. Washington, DC: National Academy Press.

IOM. 1990. *Meeting Physicians' Needs for Medical Information on Occupations and Environments*. Washington, DC: National Academy Press.

Kearney TE, Olson KR, Bero LA, Heard SE, Blanc PD. 1995. Health care costs of public use of a regional poison control center. *Western Medical Journal* 162:499–504.

Lovejoy FH Jr, Robertson WO, Woolf AD. 1994. Poison centers, poison prevention, and the pediatrician. *Pediatrics* 94:220–224.

Mullaly-Quijas P, Ward DH, Woelfl N. 1994. Using focus groups to discover health professionals' information needs: A regional marketing study. *Bulletin of the Medical Library Association* 82(3):305–311.

NEDI (National Environmental Data Index). 1996. *National Environmental Data Index* [http://esdim.noaa.gov]. November.

NLM (National Library of Medicine). 1993. *Improving Toxicology and Environmental Health Information Services*. Report of the Board of Regents Long Range Planning Panel on Toxicology and Environmental Health. NIH Publication No. 94-3486. Bethesda, MD: NLM.

NLM. 1995. *National Library of Medicine Programs and Services, 1994*. NIH Publication No. 95-256. Bethesda, MD: NLM.

Seiver R. 1989. Conducting focus group research. *Journal of College Admissions* 122:4–9.

Wallingford KT, Ruffin AB, Ginter KA, Spann ML, Johnson FE, Dutcher GA, Mehnert R, Nash DL, Bridgers JW, Lyon BJ, Siegel ER, Roderer NK. 1996. Outreach activities of the National Library of Medicine: A five-year review. *Bulletin of the Medical Library Association* 84(2 Suppl).

Glossary and Acronyms

AAPCC	American Association of Poison Control Centers
AIDS	acquired immune deficiency syndrome
AIDSDRUGS	a factual online database (available on MEDLARS) containing information on chemical and biologic agents being evaluated in AIDS clinical trials
AIDSLINE	a bibliographic online database (available on MEDLARS) covering the scientific literature on AIDS and related topics
AIDSTRIALS	a factual online database (available on MEDLARS) containing information on the clinical trials of substances being evaluated for use against AIDS, HIV infection, and AIDS-related opportunistic diseases
AOEC	Association of Occupational and Environmental Clinics
ATSDR	Agency for Toxic Substances and Disease Registry
CANCERLIT	Cancer Literature, a bibliographic online database (available on MEDLARS) covering the scientific literature on cancer epidemiology, pathology, treatment, and research

CAS	Chemical Abstracts Service, a division of the American Chemical Society; CAS is an abstracting and indexing service for the scientific literature on chemistry.
CAS Registry Number	an identification number assigned by CAS to each unique chemical entity
CCRIS	Chemical Carcinogenesis Research Information System, a factual online database (available on MEDLARS through TOXNET) sponsored by NCI containing scientifically evaluated data derived from carcinogenicity, mutagenicity, tumor promotion, and tumor inhibition tests
CDC	Centers for Disease Control and Prevention
CD-ROM	compact disk read-only memory
CERCLA	Comprehensive Environmental Response, Compensation, and Liability Act of 1980, also known as the Superfund Act
ChemID	Chemical Identification, a nonroyalty chemical nomenclature and dictionary online database (available on MEDLARS) containing information on more than 294,000 chemical substances
CHEMLEARN	a computer-based tutorial, developed by SIS, to teach librarians, information specialists, and scientists how to effectively search ChemID and CHEMLINE
CHEMLINE	Chemical Dictionary Online, a royalty chemical nomenclature and dictionary online database (available on MEDLARS) containing information on more than 1.4 million chemical substances
CRISP	Computer Retrieval of Information on Scientific Projects, an online database providing information on NIH-funded research grants

GLOSSARY AND ACRONYMS   133

DART	Developmental and Reproductive Toxicology, a bibliographic online database (available on MEDLARS through TOXNET) containing citations to the scientific literature on teratology and other aspects of developmental toxicology
DHHS	U.S. Department of Health and Human Services
DIRLINE	Directory of Information Resources Online, a directory database (available on MEDLARS) containing information on more than 17,000 health and biomedical organizations, including federal, state, and local agencies, professional societies, support groups, academic and research institutions, and voluntary organizations
DOCLINE	NLM's automated interlibrary loan request and referral system
DoT	U.S. Department of Transportation
EHPC	Environmental Health Policy Committee, U.S. Department of Health and Human Services
EINECS	European Inventory of Existing Commercial Chemical Substances
ELHILL	MEDLARS software designed for bibliographic retrieval; a gateway links ELHILL to TOXNET allowing users to search all NLM databases.
EMIC and EMICBACK	Environmental Mutagen Information Center database (and its backfile, pre-1950 to 1991), bibliographic online databases (available on MEDLARS through TOXNET) covering the scientific literature on chemical, biologic, and physical agents that have been tested for mutagenic activity
EPA	U.S. Environmental Protection Agency

ETICBACK	Environmental Teratology Information Center backfile, a bibliographic online database (available on MEDLARS through TOXNET) covering the scientific literature on teratology (pre-1950 to 1989); ETICBACK is continued by the DART database.
FDA	Food and Drug Administration
FEDRIP	Federal Research in Progress, an online database containing information on ongoing federally funded research projects in the physical sciences, engineering, and life sciences
FEMA	Federal Emergency Management Agency
FIFRA	Federal Insecticide, Fungicide, and Rodenticide Act (1947)
FTE	full-time employee
FY	fiscal year
GENE-TOX	Genetic Toxicology, a factual online database (available on MEDLARS through TOXNET) created by EPA containing peer-reviewed genetic toxicology data on approximately 3,000 chemicals
Grateful Med	an NLM-developed software package that facilitates searching of MEDLINE and other NLM online files
HBCUs	Historically Black Colleges and Universities
HIC	Health Information Center of the Wheaton Regional Library, Montgomery County, MD; NLM sponsors a project in conjunction with HIC to provide environmental health and HIV/AIDS information resources to the general public.
HIV	human immunodeficiency virus

GLOSSARY AND ACRONYMS

HSDB	Hazardous Substances Data Bank, a factual online database (available on MEDLARS through TOXNET) containing peer-reviewed toxicity, regulatory, emergency treatment, and environmental fate information on more than 4,500 chemicals
IAIMS	Integrated Advanced Information Management Systems, IAIMS grants are available through NLM to support institution-wide computer networks linking administrative, patient care, research, and education-related information systems.
IARC	International Agency for Research on Cancer
IOM	Institute of Medicine
IRIS	Integrated Risk Information System, a factual online database produced by EPA (available on MEDLARS through TOXNET) containing information on the carcinogenic and noncarcinogenic health risk assessment of more than 660 chemical substances
ISG	Information Systems Grants, grants available through NLM that provide funds for improving or expanding access to information technologies; larger hospitals and academic health centers are the primary recipients of these grants.
ISM	Information Sources Map, one of four knowledge components of the UMLS project; ISM is a database with information on the scope, location, and access points of biomedical databases.
LOANSOME DOC	a feature of the Grateful Med software program that allows health professionals to order full-text documents for references retrieved in MEDLINE and other NLM databases
Medical Informatics	the study of medical information, including medical decision making, cognitive processes, human-machine interface, knowledge representation, and information storage and retrieval

MEDLARS	Medical Literature Analysis and Retrieval System, NLM's network of more than 40 online bibliographic and factual databases
MEDLINE	Medical Literature Online, NLM's major bibliographic database covering the biomedical literature
MED75, 80, 85, 90	back files of MEDLINE
MeSH	Medical Subject Headings, NLM's controlled vocabulary thesaurus consisting of terms arranged in both an alphabetical and a hierarchical structure
Metathesaurus	a knowledge source component of UMLS that establishes conceptual links between biomedical terms from diverse source vocabularies and classifications
MTD	maximum threshold dosage
NAS	National Academy of Sciences
NCEH	National Center for Environmental Health
NCI	National Cancer Institute
NCTR	National Center for Toxicological Research
NEDI	National Environmental Data Index
NIEHS	National Institute of Environmental Health Sciences
NIH	National Institutes of Health
NIOSH	National Institute for Occupational Safety and Health
NLM	National Library of Medicine
NN/LM	National Network of Libraries of Medicine, a network of more than 4,500 health science libraries
NTIS	National Technical Information Service

NTP	National Toxicology Program
ORNL	Oak Ridge National Laboratory, a U.S. Department of Energy multiprogram laboratory
OSHA	Occupational Safety and Health Administration
PDQ	Physician's Data Query, a factual online database established by NCI containing information on cancer treatment and clinical research programs (available on MEDLARS)
POISINDEX	a commercial database produced by Micromedex, Inc., that provides toxicity and treatment information for exposures to household products and other chemical substances
R&D	research and development
RML	Regional Medical Library, major health science libraries that administer and coordinate services in the eight geographical regions of the NN/LM
RN	Chemical Abstracts Registry Number
RTECS	Registry of Toxic Effects of Chemical Substances, a factual online database produced by NIOSH (available on MEDLARS through TOXNET) that provides chemical toxicity data on more than 133,000 substances
SARA	Superfund Amendments and Reauthorization Act (1986)
SIS	Specialized Information Services, the NLM division with responsibilities for the management of the TEHIP program and AIDS-related databases
SPECIALIST	a knowledge source component of UMLS that establishes lexical entries for biomedical terms and is used in natural language processing programs

SRP	Scientific Review Panel, experts in toxicology, medicine, and the environmental sciences that regularly review the content of HSDB
Superfund Act	(see CERCLA)
SUPERLIST	a feature of the ChemID database providing information on the chemicals that appear on key state and federal regulatory and environmental health listings
TEHIP	Toxicology and Environmental Health Information Program, NLM's program that encompasses 16 factual and bibliographic databases providing toxicology and environmental information
Telnet	program that lets Internet users log into computers other than their own host computers
TIP	Toxicology Information Program, NLM program created in 1967 to disseminate toxicology information; in 1994, the program changed its name to TEHIP.
TIPCOM	Toxicology Information Program Committee, a National Academy of Sciences advisory committee that provided policy and technical advice to SIS from 1969 to 1994
TOMES	Toxicology, Occupational Medicine, and Environmental Series, a commercial database produced by Micromedex, Inc., that provides medical and hazard management information for exposures to chemical substances
TOXLINE	Toxicology Information Online, a multicomponent, non-royalty bibliographic online database (available on MEDLARS) providing citations to the worldwide literature on toxicology and related fields
TOXLINE65	TOXLINE backfile that primarily covers citations from 1965 to 1980
TOXLIT	Toxicology Literature, a royalty-based bibliographic online database (available on MEDLARS) composed primarily of citations from Chemical Abstracts

GLOSSARY AND ACRONYMS

TOXLIT65	TOXLIT backfile that primarily covers citations from 1965 to 1980
TOXNET	Toxicology Data Network, NLM's microprocessor-based system for building, maintaining, and online delivery of databases in toxicology and environmental health; a gateway links the TOXNET systems with ELHILL so that the user can search all NLM databases.
TRI	Toxic Chemical Release Inventory, a series of online databases (available on MEDLARS through TOXNET) created by EPA from industry submissions of annual environmental releases (approximately 300 chemicals)
TRIFACTS	Toxic Chemical Release Inventory Facts, an online database (available on MEDLARS through TOXNET) containing summarized information on the health, ecological effects, safety, and handling of chemicals listed in TRI
TSCA	Toxic Substance Control Act (1976)
TSCAINV	Toxic Substances Control Act Inventory
UMLS	Unified Medical Language System, a multicomponent NLM research and development initiative working on the retrieval and integration of biomedical information from a variety of sources
WWW	World Wide Web

Appendixes

A

Acknowledgments

The committee thanks the associations and organizations that assisted the committee by disseminating the questionnaire and all of the individual respondents who took the time to complete the questionnaire, provide the committee with their views, and describe their experiences (Appendix B). Additionally, the committee expresses its appreciation to the workshop participants listed in Appendix C and to those individuals listed below who provided input to the committee.

Stacey Arnesen
National Library of Medicine

William Carey
Children's Hospital of Philadelphia

Linda Clever
California Pacific Medical Center

Ann Cox
American Association of
  Occupational Health Nurses

Tamas Doszkocs
National Library of Medicine

Gale Dutcher
National Library of Medicine

Odelia Funke
Environmental Protection Agency

Jeanne Goshorn
National Library of Medicine

Larry Green
University of Colorado School of
  Medicine

Mike Hazard
National Library of Medicine

A.C. Howerton
Micromedex, Inc.

Kathy Kirkland
Association of Occupational and
  Environmental Clinics

Donald A.B. Lindberg
National Library of Medicine

Jane Lipscomb
National Institute for Occupational
  Safety and Health

Carol Maczka
National Research Council

Nina Matheson
Johns Hopkins University School
  of Medicine

Donald Mattison
University of Pittsburgh Graduate
  School of Public Health

Clifford Mitchell
Johns Hopkins University

Alan Nelson
American Society of Internal
  Medicine

Paul Pentel
American College of Medical
  Toxicology

George Rodgers
American Association of Poison
  Control Centers

Bonnie Rogers
University of North Carolina,
  Chapel Hill

David Sandler
National Research Council

Harold Schoolman
National Library of Medicine

Anthony Scialli
Georgetown University Hospital

Melvin Spann
National Library of Medicine

Dorothy Stroup
National Library of Medicine

Sandra Susten
Agency for Toxic Substances and
  Disease Registry

Diane Wagener
National Center for Health
  Statistics

Phil Wexler
National Library of Medicine

B

Questionnaire

To obtain input from a number of health professionals on their use of toxicology and environmental health information resources in general and their use of NLM databases specifically, the committee, in conjunction with health professional organizations, developed and disseminated a questionnaire. Several methods were used to distribute the questionnaire; however, the committee did not attempt to obtain a random scientific sample. The responses were viewed by the committee as indicative but not definitive.

Copies of the questionnaire along with background information on the IOM study were distributed to members of the American Association of Occupational Health Nurses, the American Association of Poison Control Centers, and the Association of Occupational and Environmental Clinics (AOEC).[1] Individuals who attended the committee's workshop (see Appendix C) also completed the questionnaire. To broaden the respondent base, the Internet was used to distribute an online version of the questionnaire to the subscribers of two Internet listservs[2] (mailing lists)—one on occupational and environmental medicine and the other sponsored by the American College of Medical Toxicology.

A total of 247 responses were received—77 online responses, 140 responses by mail, and 30 responses from the workshop. Because of the distribution of the

---

[1] Each AOEC clinic was sent three questionnaires and was asked to distribute them to a physician, a nurse, and an industrial hygienist.

[2] Listservs are subscription email networks that are focused on specific topic areas. Individuals subscribe to the listserv and then receive (and can send) email messages on the topic to other listserv members.

questionnaire on the Internet, it was not possible to estimate the number of individuals who received a copy of the questionnaire but did not respond. Responses were received from 33 states and from 3 foreign countries.

As seen in the tabulation of the responses (see below), the majority of the respondents work in the field of occupational and environmental health, function in a clinical role, and are familiar with computers. Although the respondents expressed a preference for online information (Question 12), most respondents currently consult textbooks and reference materials when locating toxicology and environmental health information (Figure B.1). Other information resources used include commercial databases, poison control centers, colleagues, and the NLM databases. The health professionals who responded to the questionnaire use toxicology and environmental health information for a variety of purposes including patient care, teaching and education, worker safety, and risk analysis (Question 15).

The questionnaire posed a series of questions that were specific to use of the TEHIP databases. As seen in the responses to Question 16, TOXLINE/TOXLIT (followed by RTECS and HSDB) were the most familiar to the respondents. When asked specifically what factors limited the use of NLM's toxicology and environmental health databases, respondents pinpointed access to the databases as the major barrier, followed by training, search language, and the front-end interface (Figure B.2). It is noteworthy that 57 of the 247 respondents were not aware of the databases. The final question asked respondents to choose the one area that, if changed or improved, would make the databases more useful and accessible. Respondents again chose access to the databases as the leading area needing improvement, followed by database training, front-end interface, and search language.

The committee gained a great deal of input through the responses to the questionnaire and used the information, in conjunction with the summaries of focus group discussions (see Appendix C), input from guest speakers, and discussions with colleagues, to inform the committee's deliberations and decision making.

APPENDIX B

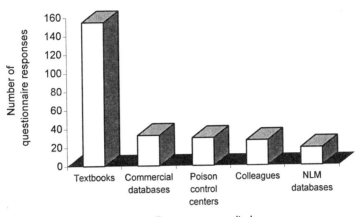

FIGURE B.1 Toxicology and environmental health information resources most often consulted.

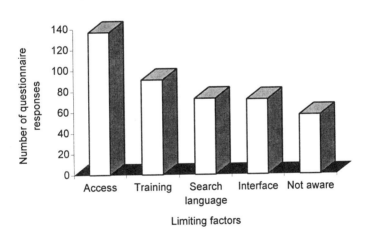

FIGURE B.2 Primary factors limiting use of the NLM toxicology and environmental health databases.

Summary of Responses

Toxicology and Environmental Health Online Databases

The National Library of Medicine's (NLM) Toxicology and Environmental Health Information Program is comprised of factual and bibliographic databases with a wide range of information including the toxicities of numerous chemicals. The Institute of Medicine (IOM), National Academy of Sciences, is currently conducting a study on strategies for increasing and improving the accessibility and availability of those NLM databases for the work of health professionals.

This survey questionnaire is designed to collect information on your awareness and use of the NLM toxicology and environmental health databases. Your input would greatly assist the IOM committee in its deliberations and would assist the National Library of Medicine in improving its Toxicology and Environmental Health Information Program. A short synopsis of each database can be found in Question 16. Thank you for your time!

1. **Age or year of completion of undergraduate education:**
    [Average age, 42.5]

2. **Please check the ONE category below that best describes the work you do.**

Emergency medicine	19	Industrial hygiene	17
Risk analysis or policy	1	Family care/primary care	7
Clinical/basic research	3	Occupational/environ. health	147
Health professional education	4	Toxicology/pharmacology	1
Library science/informatics	6	Public health	10
Community level organization	1	Poison control	12
National advocacy organization	0	Other (*please specify*)	23

3. **In what setting do you primarily work?**

Academic	111	Government	41
Private for profit	47	Private not-for-profit	30
Other (*please specify*)	17		

4. **Do you see patients or function in a clinical role?**

Yes	243
No	71
If yes: More than 20 hours per week	71
Less than 20 hours per week	81

APPENDIX B

5. **Please indicate your training, check all that apply.**
   L.P.N. 1 M.P.H. 69 M.D. 100
   CIH 3 R.N. 68 Ph.D. 33
   P.A. 4 M.S.N. 5
   Other (*please specify*) 60

6. **Do you use computers?** Yes 243

7. **How would you describe your computer skills/literacy?**
   None/Basic 33   Intermediate 174   Advanced 39

8. **Do you use the Internet?** Yes 192

9. **Do you search any online databases? Please indicate the extent of your online searching experience.**
   No online searching 61
   Medline primarily but occasionally other online databases 100
   Medline only 27
   Numerous online databases 54

10. **How often on average do you need toxicology/environmental health information?**
    Several times a day 55   Once a week 84
    Once a day 46   Once a month 50
    Once a year 9

11. **What toxicology/environmental health information resource do you most frequently use?** *Please check the ONE most often consulted.*
    Textbooks, reference materials 155
    NLM's toxicology/environmental health databases 19
    Local poison control center 30
    Commercial databases (e.g., Micromedex) 33
    Public health department 7
    Colleagues 27
    Other databases (*please specify*) 11
    Other sources (*please specify*) 10

12. **What format do you <u>prefer</u> for locating toxicological and environmental health information?**
    Text 65   Telephone 18
    CD-ROM 54   Online 98   Other (*please specify*) 5

13. **If you use databases to gather/find toxicological and environmental health information, which ones do you use?**
    Please list specific databases:
    Medline 54   TOXLINE/TOXLIT 40
    NIOSHTIC 22   TOXNET 33
    Other 34   Poisindex/TOMES 39
    RTECS 18

14. **Who searches for your toxicological and environmental health information?**

I search the databases	173	Other staff members	20
The library	25	Other (*please specify*)	13

15. **Toxicology and environmental health information can be used in many ways. Please check the main ways you use this information.** *Please check all that apply.*

Patient care	174	Teaching/education	169
Clinical research	79	Policy decisions	75
Basic research	49	Risk analysis	93
Worker safety	144	Retrieve information for others	
Community/advocacy	47	(e.g., library searching)	36
Other (*please specify*)	14		

16. **This question is on the next page and includes a description of the NLM toxicology/environmental health databases.**

17. **We are interested in identifying the factors that may limit your use of the NLM toxicology/environmental health databases.** *Please check ALL of the factors below that limit your use of the database.*

Access to the databases	137	Extent of peer review	24
Search language	73	Database training	91
Front-end interface	72	Database documentation	17
Complexity of the database records	45		
Database information content	31		
I was NOT AWARE of any of the databases.	57		
The databases have no application to my work.	5		
Other (*please specify*):	20		

18. **Please check the ONE area that if changed or improved would make the NLM toxicology/environmental health databases more useful and accessible to you.** *Please check only ONE.*

Access to the databases	91	Extent of peer review	2
Search language	21	Database training	41
Front-end interface	26	Database documentation	0
Complexity of the database records	9		
Database information content	7		
These databases are useful to me without any changes.			14
Other (*please specify*):			8

Thank you for your input! We welcome any additional comments:

16. We are interested in learning about your use of the NLM toxicology/environmental health databases and their potential usefulness in your work. For each of the databases described below, check the "YES" box if you use or have used this database. If you have not used this database, check one of the "NO" boxes to indicate whether based on the short description this database may be USEFUL or NOT USEFUL in your work.

Database	Database Information Description	YES	NO, but USEFUL	NO, NOT USEFUL
Chemical Carcinogenesis Research Information System (CCRIS)	Experimental data from carcinogenicity, mutagenicity, tumor promotion and tumor inhibition testing. Includes animal and epidemiologic studies. The database is evaluated and maintained by the National Cancer Institute.	31	152	42
Chemical Identification (ChemID)	Chemical dictionary file providing information on over 284,000 compounds; SUPERLIST data provided.	33	147	43
Chemical Dictionary Online (CHEMLINE)	Chemical dictionary file for over 1 million compounds.	47	139	37
Developmental and Reproductive Toxicology (DART) and Environmental Teratology Information Center (ETICBACK)	Bibliographic databases covering literature on teratology and other aspects of developmental and reproductive toxicology. The database is funded by EPA and NIEHS.	24	136	36
Directory of Information Resources (DIRLINE)	Directory of organizations providing biomedical information; contact information is available including address and phone number.	17	145	55
Environmental Mutagen Information Center (EMIC and EMICBACK)	Bibliographic database on chemical, biological, and physical agents tested *in vivo*, *in utero*, or *in vivo* for genotoxic activity. The database is produced by Oak Ridge National Laboratory with funding from EPA and NIEHS.	15	137	69

continues

Question 16 Continued

Database	Database Information Description	YES	NO, but USEFUL	NO, NOT USEFUL
Genetic Toxicology (GENE-TOX)	Data bank created by EPA with results from expert review of the scientific literature on chemicals tested for mutagenicity.	17	132	66
Hazardous Substances Data Bank (HSDB)	Factual database maintained by NLM focusing on the toxic effects, environmental fate, and safety and handling of hazardous chemicals. Includes human exposure, emergency medical treatment, and regulatory requirements data.	68	140	16
Integrated Risk Information System (IRIS)	EPA health risk and regulatory information on 590 chemicals; includes carcinogenic and non-carcinogenic risk assessment data.	57	137	27
Registry of Toxic Effects of Chemical Substances (RTECS)	NIOSH file of toxic effects data on over 130,000 chemicals. Both acute and chronic effects are described, including data on skin and eye irritation, carcinogenicity, mutagenicity, and reproductive consequences.	90	125	12
Toxic Chemical Release Inventory Series (TRI)	EPA files of annual estimated releases of toxic chemicals to the environment and amounts transferred to waste sites; includes information on facilities that manufacture, process, or use these chemicals	28	114	77
Toxic Chemical Release Inventory Fact Sheets (TRIFACTS)	Health, ecological effects, safety, and handling information on most chemicals listed in the TRI database.	17	150	49
Toxicology Information Online (TOXLINE and TOXLIT)	Bibliographic database with toxicology-related citations compiled from MEDLINE and 17 other sources including Chemical Abstracts, BIOSIS, and NIOSHTIC	104	99	24

C

Workshop on Toxicology and Environmental Health Information Resources: Agenda, Participants, and Summary of Focus Group Discussions

Institute of Medicine
National Academy of Sciences
Committee on Toxicology and Environmental Health Information Resources
for Health Professionals

Cecil and Ida Green Building
2001 Wisconsin Avenue, N.W.
Washington, DC
Wednesday, May 22, 1996
9:00 a.m.–2:30 p.m.

AGENDA

9:00–9:30 a.m.	Registration—Room 110
9:30–9:45 a.m.	Welcome—Room 110 Howard Kipen, Committee Chair Kathleen Stratton, Deputy Director, IOM Division of Health Promotion and Disease Prevention
9:45–12:00 p.m.	**Current Use of Toxicology/Environmental Health Information Resources** *Focus Groups*—Rooms 114, 116, 127, 128
12:00–1:00 p.m.	Lunch—Room 110/South Reception Area
1:00–2:30 p.m.	**Toxicology/Environmental Health Databases** *Demo and Discussion*—Rooms 118, 127
2:30 p.m.	Adjourn

INSTITUTE OF MEDICINE

**Workshop on
Toxicology and Environmental Health Information Resources
May 22, 1996**

PARTICIPANTS

Allison Ansher
Virginia Department of Health
Manassas, VA

Gershon Bergeisen
U.S. Environmental Protection
  Agency
Washington, DC

Karen Bolla
Johns Hopkins Bayview Medical
  Center
Baltimore, MD

Randall Brinkhuis
U.S. Environmental Protection
  Agency
Washington, DC

Gail Buckler
Robert Wood Johnson Medical
  School
Piscataway, NJ

Keith Burkhart
Central Pennsylvania Poison
  Center
Hershey, PA

Maureen Caborette
Johns Hopkins University School
  of Hygiene and Public Health
Baltimore, MD

Ann Cary
George Mason University
Fairfax, VA

Cathleen Clancy
National Capital Poison Center
Washington, DC

Robert Copeland
Howard University College of
  Medicine
Washington, DC

Mark Ennen
George Washington University
  School of Medicine
Washington, DC

Steve Galson
U.S. Department of Energy
Washington, DC

Rosemary Garrett
U.S. Postal Service
Washington, DC

Gary Greenberg
Duke University Medical Center
Durham, NC

Bryan Hardin
National Institute for Occupational
  Safety and Health
Atlanta, GA

APPENDIX C

Joseph Kaczmarczyk
Division of Federal Occupational
  Health
U.S. Public Health Service
Bethesda, MD

Linda Karbonit
Virginia Department of Health
Vienna, VA

Phyllis Lansing
University of Maryland at
  Baltimore
Baltimore, MD

Melissa McDiarmid
Occupational Health and Safety
  Administration
Washington, DC

Robert Mueller
Virginia Commonwealth
  University
Richmond, VA

Thomas Obisesan
Howard University Hospital
Washington, DC

Jerome Paulson
George Washington University
  School of Medicine
Washington, DC

Dalton Paxman
U.S. Department of Health and
  Human Services
Washington, DC

Janet Phoenix
National Safety Council
Washington, DC

Karyn Pomerantz
George Washington University
  School of Medicine
Washington, DC

Liz Ribadenyra
U.S. Department of Health and
  Human Services
Washington, DC

Diane Rodill
Division of Federal Occupational
  Health
U.S. Public Health Service
Bethesda, MD

Nadia Shalauta
Johns Hopkins University School
  of Hygiene and Public Health
Baltimore, MD

Anthony Shephard
Howard University College of
  Medicine
Washington, DC

Ginny Stone
University of Maryland at
  Baltimore
Baltimore, MD

Gregory Szlyk
George Washington University
  School of Medicine
Washington, DC

Terrance Tobias
Dalghren Memorial Library
Georgetown University
Washington, DC

Anne Tschirgi
Maryland Poison Center
Baltimore, MD

Daniel Wartenberg
Environmental and Occupational
Health Sciences Institute
Piscataway, NJ

SUMMARY OF FOCUS GROUP DISCUSSIONS

Thirty-four individuals participated in four focus groups sponsored by the IOM Committee on Toxicology and Environmental Health Information Resources for Health Professionals. The focus groups were held at the morning session of the committee's May 22, 1996 workshop. Facilitators from Jupiter Corporation, a scientific consulting firm, worked with input from committee members and IOM staff to develop a structured set of questions and issues that were then discussed in the focus group sessions.

Focus group participants included specialists in toxicology and environmental health (e.g., poison control center staff and occupational and environmental health professionals); government science and health policy advisors; researchers and information specialists; practicing health professionals in general practice or specialties other than toxicology, occupational medicine, or environmental health; health educators; and medical students.

Participants of the focus groups were asked for input on four key topics:

1. toxicology and environmental health information needs;
2. methods and sources used to locate and retrieve toxicology and environmental health information;
3. current use of toxicology and environmental health databases and other information resources in meeting information needs; and
4. future directions for toxicology and environmental health information resources.

Information Needs

The frequency of use of toxicology and environmental health databases and the need for toxicology and environmental health information varied widely among the participants, from several times a day to less than once a month. The groups that used toxicology and environmental health information most frequently (daily to weekly) were the specialists and researchers. Emergency room physicians and government policy advisors had an intermediate level of need and use; emergency room physicians frequently used a commercial CD-ROM database product (POISINDEX®) or call a poison control center. Physicians practicing in other fields, educators, and medical students needed toxicology and environmental health information least often (generally less than once a month).

Toxicology and environmental health information needs were categorized in two ways: (1) time sensitive (i.e., dependent on the critical need or time frame within which an answer or information is needed) and (2) the level of summarization or "distillation" of data desired (i.e., does the requester want original data or summarized—bottom-line—information). Participants felt that there is a correlation, specifically, an inverse relationship, between these two factors: acute information needs (such as immediate patient treatment) require a short, factual, bottom-line answer (or "distilled" information), whereas for a greater depth and amount of data, a longer amount of time can be taken to locate the comprehensive information needed.

Specialists, researchers, and government policy advisors had the broadest range of information needs, citing almost all topics related to toxicology and environmental health, including short-term and long-term effects of exposure, dose-response information, epidemiologic data, regulatory and policy requirements and background information leading to the drafting of regulations, scientific and regulatory information from the international community, personal protective information, workplace release and exposures, treatment protocols, status of research and studies, and analytic chemistry data.

Methods and Sources Used to Locate and Retrieve Information

Focus group participants indicated that the methods or sources of information they use depend primarily on whether the information need is acute (time critical) and secondarily on the level of toxicology and environmental health expertise of the person needing the information. Ease of locating and retrieving the desired information and cost were also factors in choosing sources. In general, people want to expend the least amount of time, effort, and money to get the information they need in the format they want.

Those with acute time-critical needs (e.g., emergency room physicians) and those requiring "distilled" information (e.g., general practice physicians responding to a patient's question about possible workplace exposure risks) primarily used textbooks, CD-ROM databases, and calls to experts.

Toxicology and environmental health specialists, government policy advisors, and research and information specialists were more likely to use online databases, as well as a wide range of other sources of information. These sources included printed material, other experts and colleagues, other agency sources of information, CD-ROM databases, and the Internet. The wide variety of sources used reflects the fact that these groups generally need a significant depth and breadth of data and information.

The starting point or "trigger" for a search for toxicology and environmental health information also varied widely. The most common starting point reported by all participants was the name of the chemical, although emergency room physicians stated that they often started with a drug's street name or a product's brand name and needed to "translate" this to the chemical name. Participants involved in patient care indicated that their starting point for an information query is frequently specific symptoms combined with information on the source of the exposure, such as the workplace or occupation. Exposure-related starting points (such as place of work, type of work or job, and geographic location) were also cited as common starting points by all participants.

Some of the participants in the focus groups conducted searches for information themselves; others stated that they had other staff members, such as assistants or research librarians, run database searches for them. The choice seemed to be based on (1) an ability to conduct a database search (access to the database and knowledge on how to conduct the search), (2) the time available to conduct the search, and (3) whether a high value was placed on seeing the data and information available and then "following leads" and data "trails." Specialists, researchers, and information specialists were most likely to conduct searches themselves. Policy advisors, educators, and medical students sometimes searched themselves and sometimes requested that searches be done by others. Physicians in general practice or nontoxicology-related fields were most likely to have others search for them.

Use of Online Toxicology and Environmental Health Databases

All 34 focus group participants used computers at home and at work. Since all were computer literate, lack of ability or unwillingness to use a computer were not barriers to using online databases. Toxicology and environmental health specialists, researchers, and information specialists used online databases most frequently, followed by government policy advisors.

Focus group participants indicated that the primary advantages to using online databases rather than other sources were the ability to obtain information, particularly a large amount of information rapidly; frequency of updates; and the convenience of being able to access information from any computer with a modem (office or home).

The primary barriers to locating information through the use of online databases cited by participants were (1) cost of use, particularly downloading large files and, for nonfrequent users, the amount of time required to develop a query, and (2) the time and expertise needed to locate and process information (many separate databases, different search protocols, and the large volume of information to sort through in nonstandard or difficult-to-use formats). Although convenience and accessibility from any computer with a modem were cited as advantages to using online databases, participants also noted that if a potential user did not have access to a computer with the appropriate hardware, software, and "subscription," this would be a barrier resulting in the use of other readily available information sources (e.g., textbooks, colleagues). Several participants also cited the inability to get full-text information (e.g., copies of study reports or articles) directly from or through the database as a barrier.

Focus group participants expressed three primary concerns with the content of online toxicology and environmental health databases:
1. Validity and reliability of information. They were concerned that summarized data did not present a true picture, particularly regarding the weight of evidence. In addition, some participants wanted both peer-reviewed and non-peer-reviewed information and wanted to know which was which.
2. Lack of availability of human data.
3. Inability to determine if a complete picture (all information) had been located or if a lack of information was due to (a) poor search technique, (b) information being "pulled" from the database due to controversy or while under study, or (c) a true lack of information.

Future Directions

All focus group participants expressed a desire for a more user-friendly system. Suggestions included the following:

- standardize protocols and terminology for searching;
- "one-stop shopping" (link databases so that they can be accessed with one query);
- lower costs;
- go to a "point-and-click" environment; and
- add the ability to email or otherwise transfer information directly and efficiently, and include the ability to access and download full texts of articles and studies.

Participants also had suggestions for improving health professionals' awareness of online databases including increasing training courses, particularly hands-on workshops, and online or demonstration disk tutorials; improved technical support resources, such as a toll-free telephone help-line, good help instructions, or user-friendly manuals and documentation; and advertising through articles in journals, newsletters, and other media.